高等学校建筑环境与能源应用工程专业规划教材

建筑环境与能源应用工程专业英语

王 方 张仙平 编

中国建筑工业出版社

图书在版编目（CIP）数据

建筑环境与能源应用工程专业英语/王方，张仙平编．—北京：中国建筑工业出版社，2018.12（2022.8 重印）
高等学校建筑环境与能源应用工程专业规划教材
ISBN 978-7-112-22938-3

Ⅰ．①建… Ⅱ．①王… ②张… Ⅲ．①建筑工程-环境管理-英语-高等学校-教材 Ⅳ．①TU-023

中国版本图书馆 CIP 数据核字(2018)第 259882 号

本书基于课程教学实践，结合作者国际交流合作与访问学者经历，面向国际化人才培养的专业建设需求，凝练专业课程体系核心知识点，注重暖通空调制冷行业最新技术与基础知识的映射，选取工程热力学、传热学、流体力学等专业基础课程和专业核心课程中的相关知识，形成本书的主要特色：（1）内容更新，教材中的内容均选自欧美最新原版书籍；（2）覆盖面更广，教材中共设置 17 个单元的教学内容，涉及空调、制冷、供热、通风、冷热源等方面的国际化技术和方案；（3）实用性更强，在系统深入地介绍专业知识的基础上，培养学生撰写和投稿英文学术论文，参加国际会议等技术流程。

本书可作为普通高等院校建筑环境与能源应用工程专业及相关专业学生的专业英语教材，也可供上述专业的教师、科研及工程技术人员参考。

为了更好地支持相应课程的教学，我们向采用本书作为教材的教师提供课件，有需要者可与出版社联系。

建工书院：http：//edu. cabplink. com
邮箱：jckj@cabp. com. cn 电话：（010）58337285

责任编辑：苏浩然 齐庆梅
责任校对：姜小莲

高等学校建筑环境与能源应用工程专业规划教材
建筑环境与能源应用工程专业英语
王 方 张仙平 编
*
中国建筑工业出版社出版、发行（北京海淀三里河路 9 号）
各地新华书店、建筑书店经销
北京红光制版公司制版
北京同文印刷有限责任公司印刷
*
开本：787×1092 毫米 1/16 印张：15 字数：374 千字
2018 年 12 月第一版 2022 年 8 月第二次印刷
定价：**38.00** 元
ISBN 978-7-112-22938-3
（33037）

前　言

在中国经济高速发展的形势下，中国暖通空调行业得到全面技术革新，逐步走进国际领先行列。在专业人才培养的过程中，国际间学生交流访学和联合培养日益频繁，国内外专家学者高端学术讲座走进大学课堂的机会日益增多，暖通空调专业教学呈现出越来越强烈的国际化趋势和需求。大批暖通空调领军企业走向国际市场，面对国外先进技术和设备，需要大量既懂专业又在英语阅读、交流、写作等方面得心应手的专门人才。为适应社会需求，国内高校建筑环境与能源应用工程专业普遍开设专业英语课程，受到各校专业发展历程和专业人才培养特色的影响，专业英语教材的内容体系差异较大，新时代对建筑环境与能源应用工程专业英语内容体系也提出了新要求。

本教材点主要特色体现在：(1) 内容更新，教材中的内容均选自欧美最新原版书籍；(2) 覆盖面更广，教材中共设置 17 共单元的教学内容，涉及空调、制冷、供热、通风、冷热源等方面最全面的国际化技术和方案；(3) 实用性更强，在系统深入的介绍专业知识的基础上，培养学生撰写和投稿英文学术论文，参加国际会议等技术流程。

全书综合建筑环境与能源应用工程专业课程核心知识点，共设置 17 个学习单元，其中：第 1、5、6、14、17 单元由王方（中原工学院）负责编写，第 4、7、11、12、13 单元由张仙平（河南工程学院）负责编写，第 2、3、9、10 单元由李志强（中原工学院）负责编写，第 8、15、16 单元由段焕林（河南工程学院）负责编写。

加拿大卡尔顿大学 Junjie Gu 教授审阅了全书。此外，本次编写得到了中国建筑工业出版社齐庆梅编审和苏浩然编辑的大力支持，在此一并表示诚挚的谢意。

编者水平所限，书中难免有疏漏与不妥之处，敬请广大师生指正。

CONTENTS

Unit 1　Engineering Thermodynamics

Engineers use principles drawn from **thermodynamics** and other engineering sciences, including fluid mechanics plus heat and mass transfer, to analyze and design devices intended to meet human needs. Throughout the twentieth century, engineering applications of thermodynamics helped pave the way for significant improvements in our quality of life with advances in major areas such as surface transportation, air travel, space flight, electricity generation and transmission, building heating and cooling, and improved medical practices. The wide realm of these applications is suggested by Table1-1.

Selected areas of application of engineering thermodynamics　　**Table 1-1**

Aircraft and rocket propulsion
Alternative energy systems
Fuel cells
Geothermal systems
Magnetohydrodynamic (MHD) converters
Ocean thermal, wave, and tidal power generation
Solar-activated heating, cooling, and power generation
Thermoelectric and thermionic devices
Wind turbines
Automobile engines
Bioengineering applications
Biomedical applications
Combustion systems
Compressors, pumps
Cooling of electronic equipment
Cryogenic systems, gas separation, and liquefaction
Fossil and nuclear-fueled power stations
Heating, ventilating, and air-conditioning systems
Absorption refrigeration and heat pumps
Vapor-compression refrigeration and heat pumps
Steam and gas turbines
Power production
Propulsion

In the twenty-first century, engineers are creating the technologies to achieve a sustainable future. Thermodynamics will continue to advance human well-being by addressing looming societal challenges due to declining supplies of energy resources: oil, natural gas, coal, and fissionable material; effects of global climate change; and growing population. Life style is expected to change in several important respects by mid-century. In the area of power use, for example, electricity will play an even greater role than that for today.

If this vision of mid-century life is correct, it will be necessary to evolve quickly from our present energy postures. As was the case in the twentieth century, thermodynamics will contribute significantly to meeting the challenges of the twenty-first century, including using fossil fuels more effectively, advancing renewable energy technologies, and developing more energy-efficient transportation systems, buildings, and industrial practices. Thermodynamics also will play a role in mitigating global climate change, air pollution, and water pollution. Applications will also be observed in bioengineering, biomedical systems, and the deployment of nanotechnology.

1.1 Concepts and Definitions

The initial step in any engineering analysis is to describe precisely what is being studied. In mechanics, if the motion of a body is to be determined, normally the first step is to define a free body and identify all the forces exerted on it by other bodies. Newton's second law of motion is then applied. In thermodynamics the term system is used to identify the subject of study. Once the system is defined and the relevant interactions with other systems are identified, one or more physical laws or relations can be applied.

The **system** is whatever we want to study. It may be as simple as a free body or as complex as an entire chemical refinery. We may want to study a quantity of matter contained within a closed, rigid-walled tank, or we may want to consider something such as a piece of pipeline through which natural gas flows. The composition of the matter inside the system may be fixed or may be changing through chemical or nuclear reactions. The shape or volume of the system being analyzed is not necessarily constant, as when a gas in a cylinder is compressed by a piston or a balloon is inflated.

Everything external to the system is considered to be part of the system's surrounding. The system is distinguished from its surroundings by a specified boundary, which may be at rest or in motion. You will see that the interactions between a system and its surroundings, which take place across the boundary, and play an important part in engineering thermodynamics.

A closed system is defined when a particular quantity of matter is under study. A closed system always contains the same matter. There can be no transfer of mass across its boundary. A special type of closed system that does not interact in any way with its surroundings is called an isolated system.

Figure 1-1 shows a gas in a piston - cylinder assembly. When the valves are closed, we can consider the gas to be a closed system. The boundary lies just inside the piston and cylinder walls, as shown by the dashed lines on the figure. Since the portion of the boundary between the gas and the piston moves with the piston, the system volume varies. No mass would cross this or any other part of the boundary. If combustion occurs, the composition of the system changes as the initial combustible mixture.

In subsequent sections of this book, we perform thermodynamic analyses of devices such as turbines and pumps through which mass flows. These analyses can be conducted in principle by studying a particular quantity of matter, a closed system, as it passes through the device. In most cases, however, it is simpler to think instead in terms of a given region of space through which mass flows. With this approach, a region within a prescribed boundary is studied. The region is called a control volume. Mass crosses the boundary of a control volume.

The system boundary should be delineated carefully before proceeding with any thermodynamic analysis. However, the same physical phenomena often can be analyzed in terms of alternative choices of the system, boundary, and surroundings. The choice of a particular boundary defining a particular system depends heavily on the convenience which allows in the subsequent analysis.

In general, the choice of system boundary is governed by two considerations: (1) what is known about a possible system, particularly at its boundaries, and (2) the objective of the analysis.

For example, Figure 1-2 shows a sketch of an air compressor connected to a storage tank. The system boundary shown on the figure encloses the compressor, tank, and all of the piping. This boundary might be selected if the electrical power input is known, and the objective of the analysis is to determine how long the compressor must operate for the pressure in the tank to rise to a specified value. Since mass crosses the boundary, the system would be a control volume. A control volume enclosing only the compressor might be chosen if the condition of the air entering and exiting the compressor were known, and the objective is to determine the electric power input.

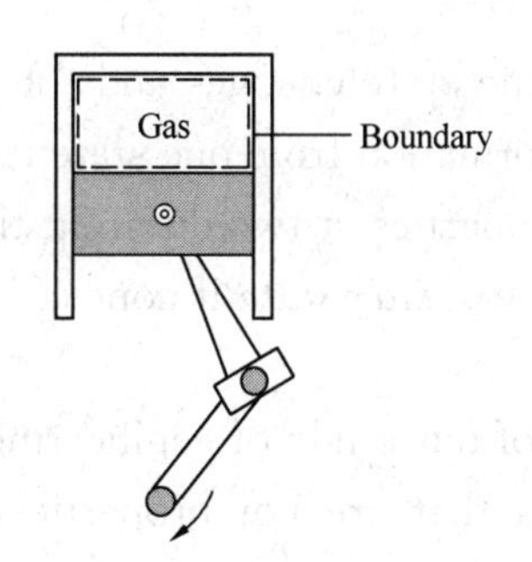

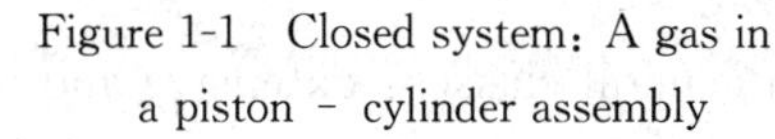
Figure 1-1 Closed system: A gas in a piston - cylinder assembly

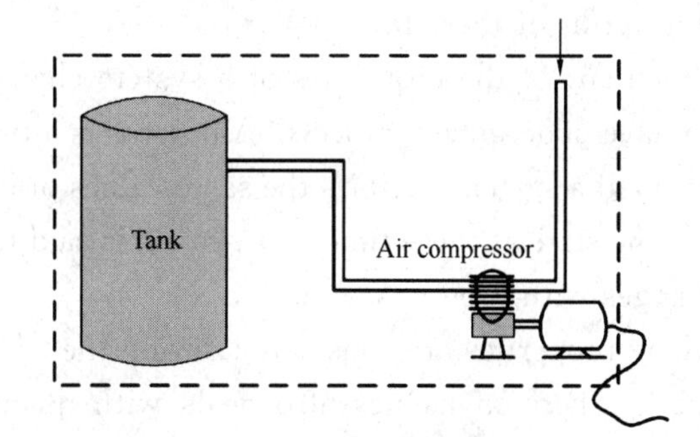

Figure 1-2 Air compressor and storage tank.

Engineers are interested in studying systems and how they interact with their surroundings. Therefore, we introduce several terms and concepts used to describe systems and how they behave.

Systems can be studied from a macroscopic or a microscopic point of view. The macroscopic approach to thermodynamics is concerned with the gross or overall behavior. This is sometimes called *classical* thermodynamics. No model of the structure of matter at the molecular, atomic, and subatomic levels is directly used in classical thermodynamics. Although the behavior of systems is affected by molecular structure, classical thermodynamics allows important aspects of system behavior to be evaluated from observations of the overall system. The microscopic approach to thermodynamics, known as *statistical thermodynamics*, is concerned directly with the structure of matter. The objective of statistical thermodynamics is to characterize by statistical means the average behavior of the particles making up a system of interest and relate this information to the observed macroscopic behavior of the system. For applications involving lasers, plasmas, high-speed gas flows, chemical kinetics, very low temperatures (cryogenics), and others, the methods of statistical thermodynamics are essential. Moreover, the microscopic approach is instrumental in developing certain data, for example *ideal gas specific heats*. For a wide range of engineering applications, classical thermodynamics not only provides a considerably more direct approach for analysis and design but also requires far fewer mathematical complications. For these reasons the macroscopic viewpoint is the one adopted in this book. Finally, relativity effects are not significant for the systems under consideration in this text.

To describe a system and predict its behavior requires knowledge of its properties and how those properties are related. A property is a macroscopic characteristic of a system such as mass, volume, energy, pressure, and temperature to which a numerical value can be assigned at a given time without knowledge of the previous behavior (*history*) of the system.

The word *state* refers to the condition of a system as described by its properties. Since there are normally relations among the properties of a system, the state often can be specified by providing the values of a subset of the properties. All other properties can be determined in terms of these few.

When any of the properties of a system change, the state changes and the system is said to have undergone a *process*. A process is a transformation from one state to another. However, if a system exhibits the same values of its properties at two different times, it is in the same state at these times. A system is said to be at *steady state* if none of its properties changes with time.

Many properties are considered during the course of our study of engineering thermodynamics. Thermodynamics also deals with quantities that are not properties, such as mass flow rates and energy transfers by work and heat.

Thermodynamic properties can be placed in two general classes: extensive and intensive. A property is called *extensive* if its value for an overall system is the sum of its values for the parts into which the system is divided. Mass, volume, energy, and several other properties introduced later are extensive. Extensive properties depend on the size or extent of a system. The extensive properties of a system can change with time, and many thermodynamic analyses consist mainly of carefully accounting for changes in extensive properties such as mass and energy as a system interacts with its surroundings.

Intensive properties are not additive in the sense previously considered. Their values are independent of the size or extent of a system and may vary from place to place within the system at any moment. Thus, intensive properties may be functions of both position and time, whereas extensive properties vary at most with time. *Specific volume*, *pressure*, and *temperature* are important intensive properties.

Classical thermodynamics places primary emphasis on equilibrium states and changes from one equilibrium state to another. Thus, the concept of *equilibrium* is fundamental. In mechanics, equilibrium means a condition of balance maintained by an equality of opposing forces. In thermodynamics, the concept is more far-reaching, including not only a balance of forces but also a balance of other influences. Each kind of influence refers to a particular aspect of thermodynamic, or complete, equilibrium. Accordingly, several types of equilibrium must exist individually to fulfill the condition of complete equilibrium; among these are mechanical, thermal, phase, and chemical equilibrium.

Criteria for these four types of equilibrium are considered in subsequent discussions. For the present, we may think of testing to see if a system is in thermodynamic equilibrium by the following procedure: isolate the system from its surroundings and watch for changes in its observable properties. If there are no changes, we conclude that the system was in equilibrium at the moment it was isolated. The system can be said to be at an *equilibrium state*.

When a system is isolated, it does not interact with its surroundings; however, its state can change as a consequence of spontaneous events occurring internally as its intensive properties, such as temperature and pressure, tend toward uniform values. When all such changes cease, the system is in equilibrium. Hence, for a system to be in equilibrium it must be a single phase or consist of a number of phases that have no tendency to change their conditions when the overall system is isolated from its surroundings. At equilibrium, temperature is uniform throughout the system. Also, pressure can be regarded as uniform throughout as long as the effect of gravity is not significant; otherwise a pressure variation can exist, as in a vertical column of liquid.

It is not necessary that a system undergoing a process be in equilibrium *during* the process. Some or all of the intervening states may be nonequilibrium states. For many such processes, we are limited to knowing the state before the process occurs and the state after the process is completed.

1.2 The First Law of Thermodynamics

Energy is a fundamental concept of thermodynamics and one of the most significant aspects of engineering analysis. In this section we discuss energy and develop equations for applying the principle of conservation of energy, the First law of Thermodynamics. The current presentation is limited to closed systems.

Energy is the propery that can be transferred to an object in order to perform work on, or to heat the object. A basic idea is that energy can be *stored* within systems in various forms. Energy also can be converted from one form to another and transferred between systems. For closed systems, energy can be transferred by *work* and *heat* transfer. The total amount of energy is *conserved* in all conversions and transfers.

Building on the contributions of Galileo and others, Newton formulated a general description of the motions of objects under the influence of applied forces. Newton's laws of motion, which provide the basis for classical mechanics, led to the concepts of work, kinetic energy, and potential energy, and these led eventually to a broadened concept of energy. The present discussion begins with an application of Newton's second law of motion.

The curved line in Figure 1-3 represents the path of a body of mass m (a closed system) moving relative to the x-y coordinate frame shown. The velocity of the center of mass of the body is denoted by $\vec{V}$. The body is acted on by a resultant force $\vec{F}$, which may vary in magnitude from location to location along the path. The resultant force is resolved into a component $\vec{F}_s$ along the path and a component $\vec{F}_n$ normal to the path. The effect of the kinetic energy component 2 $\vec{F}_s$ is to change the magnitude of the velocity, whereas the effect of the component $\vec{F}_n$ is to change the direction of the velocity. As shown in Figure 1-3, s is the instantaneous position of the body measured along the path from some fixed point denoted by 0. Since the magnitude of $\vec{F}$ can vary from location to location along the path, the magnitudes of $\vec{F}_s$ and $\vec{F}_n$ are, in general, functions of s.

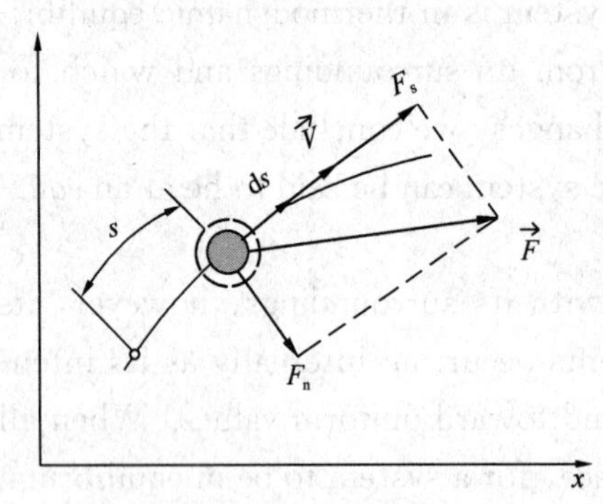

Figure 1-3 Forces acting on a moving system.

Let us consider the body as it moves from $s = s_1$, where the magnitude of its velocity is $\vec{V}_1$, to $s = s_2$, where its velocity is $\vec{V}_2$. Assume for the present discussion that the only interaction between the body and its surroundings involves the force $\vec{F}$. By Newton's second law of motion, the magnitude of the component $\vec{F}_s$ is related to the change in the magnitude of $\vec{V}$ by

$$\vec{F}_s = m\frac{d\vec{V}}{dt} \tag{1-1}$$

Using the chain rule, this can be written as

$$\vec{F}_s = m\frac{d\vec{V}}{dt} = m\frac{d\vec{V}}{ds}\frac{ds}{dt} = m\vec{V}\frac{d\vec{V}}{ds} \tag{1-2}$$

where $\vec{V}$——ds/dt. Rearranging Equation 1-2 and integrating from s_1 to s_2 gives

$$\int_{V_1}^{V_2} m\vec{V}d\vec{V} = \int_{s_1}^{s_2} \vec{F}_s ds \tag{1-3}$$

The integral on the left of Equation 1-3 is evaluated as follows

$$\int_{V_1}^{V_2} m\vec{V}d\vec{V} = \frac{1}{2}mV^2 \Big|_{V_1}^{V_2} = \frac{1}{2}m(V_2^2 - V_1^2) \tag{1-4}$$

The quantity $1/2mV^2$ is the *kinetic energy*(*KE*) of the body. Kinetic energy is a scalar quantity. The change in kinetic energy, ΔKE, of the body is

$$\Delta KE = KE_2 - KE_1 = \frac{1}{2}m(V_2^2 - V_1^2) \tag{1-5}$$

The integral on the right of Equation 1-3 is the *work* of the force F_s as the body moves from s_1 to s_2 along the path. Work is also a scalar quantity.

With Equation 1-4, Equation 1-3 becomes

$$\frac{1}{2}m(V_2^2 - V_1^2) = \int_{s_1}^{s_2} \vec{F} \, d\vec{s} \tag{1-6}$$

where the expression for work has been written in terms of the scalar product (dot product) of the force vector $\vec{F}$and the displacement vector $d\vec{s}$. Equation 1-6 states that the work of the resultant force on the body equals the change in its kinetic energy. When the body is accelerated by the resultant force, the work done on the body can be considered a *transfer* of energy *to* the body, where it is *stored* as kinetic energy.

Kinetic energy can be assigned a value knowing only the mass of the body and the magnitude of its instantaneous velocity relative to a specified coordinate frame, without regard for how this velocity was attained. Hence, kinetic energy is a property of the body. Since kinetic energy is associated with the body as a whole, it is an extensive property.

Equation 1-6 is the principal result of the previous section. Derived from Newton's second law, the equation gives a relationship between two *defined* concepts: kinetic energy and work. In this section it is used as a point of departure to extend the concept of energy. To begin, refer to Figure2-2, which shows a body of mass m that moves vertically from an elevation z_1 to an elevation z_2 relative to the surface of the earth. Two forces are shown acting on the system: a downward force due to gravity with magnitude mg and a vertical force with magnitude R representing the resultant of all *other* forces acting on the system.

The work of each force acting on the body shown in Figure 1-4 can be determined by using the definition previously given. The total work is the algebraic sum of these individual values. In accordance with Equation 1-6, the total work equals the change in kinetic energy. That is

$$\frac{1}{2}m(V_2^2 - V_1^2) = \int_{z_1}^{z_2} R \, dz - \int_{z_1}^{z_2} mg \, dz \tag{1-7}$$

A minus sign is introduced before the second term on the right because the gravitational force is directed downward and z is taken as positive upward.

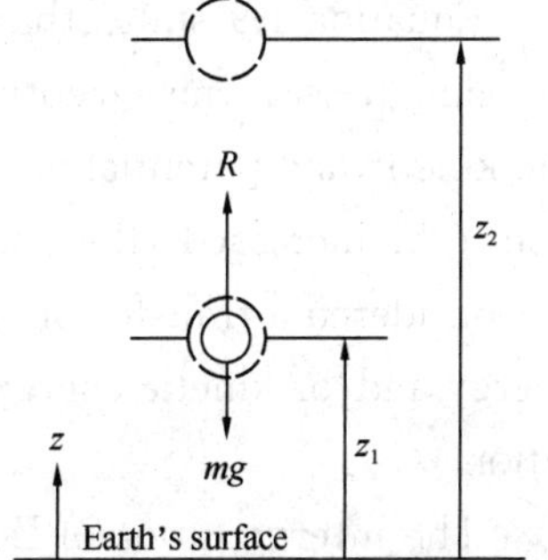

Figure 1-4 Illustration used to introduce the potential energy concept.

The first integral on the right of Equation 1-7 represents the work done by the force $\vec{R}$ on the body as it moves vertically from z_1 to z_2. The second integral can be evaluated as follows

$$\frac{1}{2}m(V_2^2 - V_1^2) + mg(z_2 - z_1) = \int_{z_1}^{z_2} R\mathrm{d}z \tag{1-8}$$

where the acceleration of gravity has been assumed to be constant with elevation. By incorporating Equation 1-8 into Equation 1-7 and rearranging

$$\int_{z_1}^{z_2} mg\,\mathrm{d}z = mg(z_2 - z_1) \tag{1-9}$$

The quantity mgz is the *gravitational potential energy* (PE). The change in gravitational potential energy, ΔPE, is

$$\Delta PE = PE_2 - PE_1 = mg(z_2 - z_1) \tag{1-10}$$

Potential energy is associated with the force of gravity and is therefore an attribute of a system consisting of the body and the earth together. However, evaluating the force of gravity as mg enables the gravitational potential energy to be determined for a specified value of g knowing only the mass of the body and its elevation. With this view, potential energy is regarded as an extensive property of the body. Throughout this book it is assumed that elevation differences are small enough that the gravitational force can be considered constant. The concept of gravitational potential energy can be formulated to account for the variation of the gravitational force with elevation, however.

To assign a value to the kinetic energy or the potential energy of a system, it is necessary to assume a datum and specify a value for the quantity at the datum. Values of kinetic and potential energy are then determined relative to this arbitrary choice of datum and reference value. However, since only *changes* in kinetic and potential energy between two states are required, these arbitrary reference specifications cancel.

Work has units of forcetimes distance. The units of kinetic energy and potential energy are the same as for work. In SI, the energy unit is the newton-meter, N • m, called the joule, J. In this book it is convenient to use the kilojoule, kJ. Commonly used English units for work, kinetic energy, and potential energy are the foot-pound force, ft • lbf, and the British thermal unit, Btu.

When a system undergoes a process where there are changes in kinetic and potential energy, special care is required to obtain a consistent set of units.

Equation 1-9 states that the total work of all forces acting on the body from the surroundings, with the exception of the gravitational force, equals the sum of the changes in the kinetic and potential energies of the body. When the resultant force causes the elevation to be increased, the body to be accelerated, or both, the work done by the force can be considered a transfer of energy to the body, where it is stored as gravitational potential energy and/or kinetic energy. The notion that energy is conserved underlies this interpretation.

The interpretation of Equation 1-9 as an expression of the conservation of energy principle can be reinforced by considering the special case of a body on which the only force acting is that due to gravity, for then the right side of the equation vanishes and the equation reduces to

$$\frac{1}{2}m(V_2^2 - V_1^2) + mg(z_2 - z_1) = 0$$

or

$$\frac{1}{2}mV_2^2 + mgz_2 = \frac{1}{2}mV_1^2 + mgz_1 \tag{1-11}$$

Under these conditions, the *sum* of the kinetic and gravitational potential energies *remains constant*. Equation 1-11 also illustrates that energy can be converted from one form to another: for an object falling under the influence of gravity only, the potential energy would decrease as the kinetic energy increases by an equal amount.

The presentation thus far has centered on systems for which applied forces affect only their overall velocity and position. However, systems of engineering interest normally interact with their surroundings in more complicated ways, with changes in other properties as well. To analyze such systems, the concepts of kinetic and potential energy alone do not suffice, nor does the rudimentary conservation of energy principle introduced in this section. In thermodynamics the concept of energy is broadened to account for other observed changes, and the principle of conservation of energy is extended to include a wide variety of ways in which systems interact with their surroundings. The basis for such generalizations is experimental evidence.

1.3 The Second Law of Thermodynamics

The objectives of the present section are to (1) motivate the need for and the usefulness of the second law, and to (2) introduce statements of the second law that serve as the point of departure for its application.

It is a matter of everyday experience that there is a definite direction for spontaneous processes. This can be brought out by considering the three systems pictured in Figure 1-5.

For system a, an object at an elevated temperatureT_i placed in contact with atmospheric air at temperature T_0 would eventually cool to the temperature of its much larger surroundings, as illustrated in Figure 1-5*a*. In conformity with the conservation of energy principle, the decrease in internal energy of the body would appear as an increase in the internal energy of the surroundings. The inverse process would not take place spontaneously, even though energy could be conserved: The internal energy of the surroundings would not decrease spontaneously while the body warmed from T_0 to its initial temperature.

For system b, air held at a high pressure p_i in a closed tank would flow spontaneously to the lower pressure surroundings at p_0 if the interconnecting valve were opened, as illustrated in Figure 1-5*b*. Eventually fluid motions would cease and all of the air would beat the same pressure as the surroundings. Drawing on experience, it should be clear that the inverse process would not take place spontaneously, even though energy could be conserved: Air would not flow spontaneously from the surroundings at p_0 into the tank, returning the pressure to its initial value.

For system c, a mass suspended by a cable at elevation z_i would fall when released, as

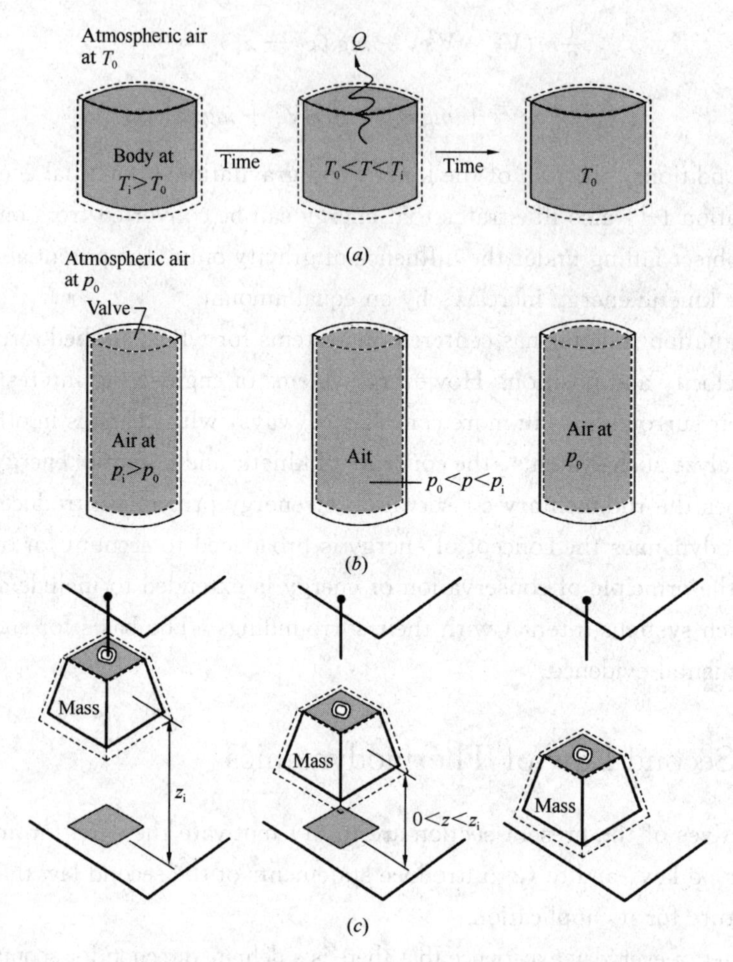

Figure 1-5 Illustrations of spontaneous processes and the eventual attainment of equilibrium with the surroundings.

(*a*) Spontaneous heat transfer; (*b*) Spontaneous expansion; (*c*) Falling mass

illustrated in Figure 1-5*c*. When it comes to rest, the potential energy of the mass in its initial condition would appear as an increase in the internal energy of the mass and its surroundings, in accordance with the conservation of energy principle. Eventually, the mass also would come to the temperature of its much larger surroundings. The inverse process would not take place spontaneously, even though energy could be conserved: The mass would not return spontaneously to its initial elevation while its internal energy or that of its surroundings decreased.

In each case considered, the initial condition of the system can be restored, but not in a spontaneous process. Some auxiliary devices would be required. By such auxiliary means the object could be reheated to its initial temperature, the air could be returned to the tank and restored to its initial pressure, and the mass could be lifted to its initial height. Also in each case, a fuel or electrical input normally would be required for the auxiliary devices to function, so a permanent change in the condition of the surroundings would result.

The foregoing discussion indicates that not every process consistent with the principle of energy conservation can occur. Generally, an energy balance alone neither enables the preferred direction to be predicted nor permits the processes that can occur to be distinguished from those that cannot. In elementary cases such as the ones considered, experience can be drawn upon to deduce whether particular spontaneous processes occur and to deduce their directions. For more complex cases, where experience is lacking or uncertain, a guiding principle would be helpful. This is provided by the Second Law.

The foregoing discussion also indicates that when left to themselves, systems tend to undergo spontaneous changes until a condition of equilibrium is achieved, both internally and with their surroundings. In some cases equilibrium is reached quickly, and in others it is achieved slowly. For example, some chemical reactions reach equilibrium in fractions of seconds; an ice cube requires a few minutes to melt; and it may take years for an iron bar to rust away. Whether the process is rapid or slow, it must of course satisfy conservation of energy. However, that alone would be insufficient for determining the final equilibrium state. Another general principle is required. This is provided by the Second Law.

By exploiting the spontaneous processes shown in Figure 1-5, it is possible, in principle, for work to be developed as equilibrium is attained. For example, instead of permitting the body of Figure 1-5*a* to cool spontaneously with no other result, energy could be delivered by heat transfer to a system undergoing a power cycle that would develop a net amount of work. Once the object attained equilibrium with the surroundings, the process would cease. Although there is an opportunity for developing work in this case, the opportunity would be wasted if the body were permitted to cool without developing any work. In the case of Figure 1-5*b*, instead of permitting the air to expand aimlessly into the lower-pressure surroundings, the stream could be passed through a turbine and work could be developed. Accordingly, in this case there is also a possibility for developing work that would not be exploited in an uncontrolled process. In the case of Figure 1-5*c*, instead of permitting the mass to fall in an uncontrolled way, it could be lowered gradually while turning a wheel, lifting another mass, and so on.

These considerations can be summarized by noting that when an imbalance exists between two systems, there is an opportunity for developing work that would be irrevocably lost if the systems were allowed to come into equilibrium in an uncontrolled way. Recognizing this possibility for work, we can pose two questions: (1) What is the theoretical maximum value for the work that could be obtained? (2) What are the factors that would preclude the realization of the maximum value? That there should be a maximum value is fully in accord with experience, for if it were possible to develop unlimited work, nobody would concern about fuel supplies and energy crisis. Also in accord with experience is the idea that even the best devices would be subject to factors such as friction that would preclude the attainment of the theoretical maximum work. The second law of thermodynamics provides the means for determining the theoretical maximum and evaluating quantitatively the factors that preclude attaining the maximum.

The second law and deductions from it are useful because they provide means for: (1) predicting the direction of processes; (2) establishing conditions for equilibrium; (3) determining the best theoretical performance of cycles, engines, and other devices; (4) evaluating quantitatively the factors that preclude the attainment of the best theoretical performance level. Additional uses of *the Second Law* include its roles in (5) defining a temperature scale independent of the properties of any thermometric substance; (6) developing means for evaluating properties such as u and h in terms of properties that are more readily obtained experimentally. Scientists and engineers have found many additional applications of the second law and deductions from it. It also has been used in economics, philosophy, and other areas far removed from engineering thermodynamics.

The six points listed can be thought as aspects of *the Second Law of Thermodynamics* and notas independent and unrelated ideas. Nonetheless, given the variety of these topic areas, it is easy to understand why there is no single statement of *the Second Law* that brings out each one clearly. There are several alternative, yet equivalent, formulations of the second law.

There are different equivalent statements of *the Second Law* as a point of departure for our study of *the Second Law* and its consequences. Although the exact relationship of these particular formulations to each of *the Second Law* aspects listed above may not be immediately apparent, all aspects listed can be obtained by deduction from these formulations or their corollaries. It is important to add that in every instance where a consequence of the Second Law has been tested directly or indirectly by experiment, it has been unfailingly verified. Accordingly, the basis of *the Second Law* of thermodynamics, like every other physical law, is experimental evidence.

Among many alternative statements of *the Second Law*, two are frequently used in engineering thermodynamics. They are the *Clausius and Kelvin-Planck statements*.

Clausius Statement of the Second Law

The Clausius statement of *the Second Law* asserts that: ***It is impossible for any system to operate in such a way that the sole result would be an energy transfer by heat from a cooler to a hotter body.*** The Clausius statement has been selected as a point of departure for the study of *the Second Law* and its consequences because it is in accord with experience and therefore easy to accept. The Kelvin-Planck statement has the advantage that it provides an effective means for bringing out important second law deductions related to systems undergoing thermodynamic cycles. One of these deductions, the Clausius inequality, leads directly to the property entropy and to formulations of the second law convenient for the analysis of closed systems and control volumes as they undergo processes that are not necessarily cycles.

The Clausius statement does not rule out the possibility of transferring energy by heat from a cooler body to a hotter body, as shown in Figure 1-6, for this is exactly what refrigerators and heat pumps accomplish. However, as the words" sole result" in the statement suggest, when a heat transfer from a cooler body to a hotter body occurs, there must be some *other effect* within the system accomplishing the heat transfer, its surroundings, or both. If the system operates in a thermodynamic cycle, its initial state is restored after each cycle, so the only place that must be examined for such *other* effects is its surroundings. For example, cooling of food is accomplished by refrigerators driven by electric motors requiring work from their surroundings to operate. The Clausius statement implies that it is impossible to construct a refrigeration cycle that operates without an input of work.

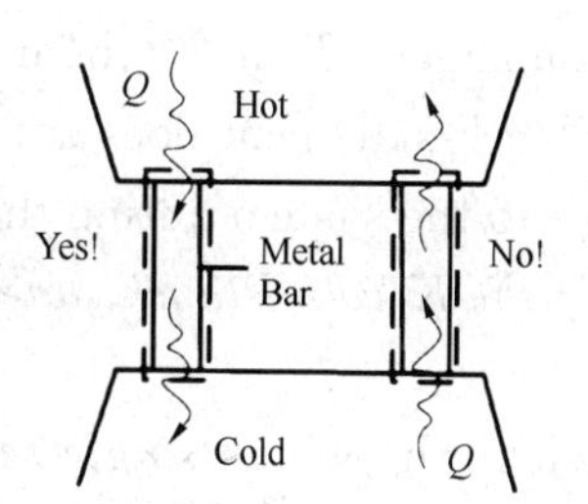

Figure 1-6 Illustration used to demonstrate the Clausius statement of the second law.

Kelvin-Planck Statement of the Second Law

Before giving the Kelvin-Planck statement of the second law, the concept of a *thermal reservoir* is introduced. A thermal reservoir, or simply a reservoir, is a special kind of system that always remains at constant temperature even though energy is added or removed by heat transfer. A reservoir is an idealization of course, but such a system can be approximated in a number of ways-by the earth's atmosphere, large bodies of water (lakes, oceans), a large block of copper, and a system consisting of two phases (although the ratio of the masses of the two phases changes as the system is heated or cooled at constant pressure, the temperature remains constant as long as both phases coexist). Extensive properties of a thermal reservoir such as internal energy can change in interactions with other systems even though the reservoir temperature remains constant.

Having introduced the thermal reservoir concept, we give the Kelvin-Planck statement of the second law: ***It is impossible for any system to operate in a thermodynamic cycle and deliver a net amount of energy by work to its surroundings while receiving energy by heat transfer from a single thermal reservoir.*** The Kelvin-Planck statement, as shown in Figure 1-7, does not rule out the possibility of a system developing a net amount of work from a heat transfer drawn from a single reservoir. It only denies this possibility if the system undergoes a thermodynamic cycle.

The Kelvin-Planck statement can be expressed analytically. To develop this, let us study a system undergoing a cycle while exchanging energy by heat transfer with a *single reservoir*. The first and second laws each impose constraints:

A constraint is imposed by the first law on the net work and heat transfer between the system and its surroundings. According to the cycle energy balance

$$W_{cycle} = Q_{cycle} \tag{1-12}$$

In words, the net work done by the system undergoing a cycle equals the net heat transfer to the system. Although the cycle energy balance allows the net work W_{cycle} be positive or negative, the second law imposes a constraint on its direction, as considered next.

According to the Kelvin-Planck statement, a system undergoing a cycle while communicating thermally with a single reservoir *cannot* deliver a net amount of work to its surroundings. That is, the net work of the cycle *can not* be positive. However, the Kelvin-Planck statement does not rule out the possibility that there is a net work transfer of energy to the system during the cycle or that the net work is zero. Thus, the *analytical form of the Kelvin-Planck statement* is

$$W_{cycle} \leqslant 0 \quad \text{(single reservoir)} \tag{1-13}$$

where the words *single reservoir* are added to emphasize that the system communicates thermally only with a single reservoir as it executes the cycle.

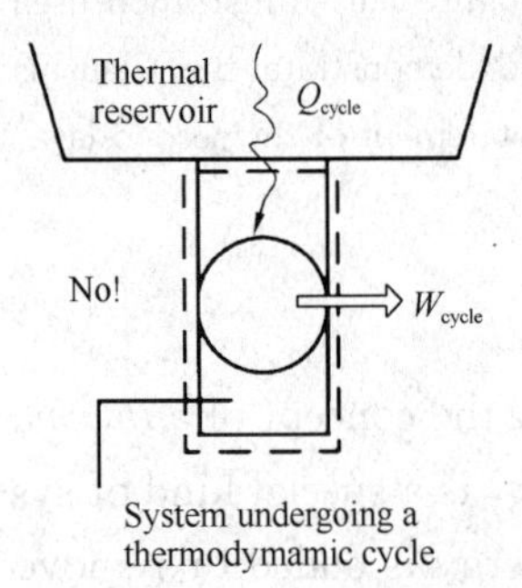

Figure 1-7 Illustration used to demonstrate the Kelvin-Planck statement of the second law.

The equivalence of the Clausius and Kelvin-Planck statements is demonstrated by showing that the violation of each statement implies the violation of the other. That a violation of the Clausius statement implies a violation of the Kelvin-Planck statement is readily shown using Figure 1-7, which pictures a hot reservoir, a cold reservoir, and two systems. The system on the left transfers energy Q_C from the cold reservoir to the hot reservoir by heat transfer without other effects occurring and thus *violates the Clausius statement*. The system on the right operates in a cycle while receiving Q_H (greater than Q_C) from the hot reservoir, rejecting Q_C to the cold reservoir, and delivering work W_{cycle} to the surroundings. The energy flows labeled on Figure 1-8 are in the directions indicated by the arrows.

Consider the combined system shown by a dotted line on Figure 1-6, which consists of the cold reservoir and the two devices. The combined system can be regarded as executing a cycle because one part undergoes a cycle and the other two parts experience no net change in their conditions. Moreover, the combined system receives energy (Q_H- Q_C) by heat transfer from a single reservoir, the hot reservoir, and produces an equivalent amount of work. Accordingly, the combined system violates the Kelvin-Planck statement. Thus, a violation of the Clausius statement implies a violation of the Kelvin-Planck statement. The equivalence of the two second-law statements is demonstrated completely when it is also shown that a violation of the Kelvin-Planck statement implies a violation of the Clausius statement. This is left as an exercise.

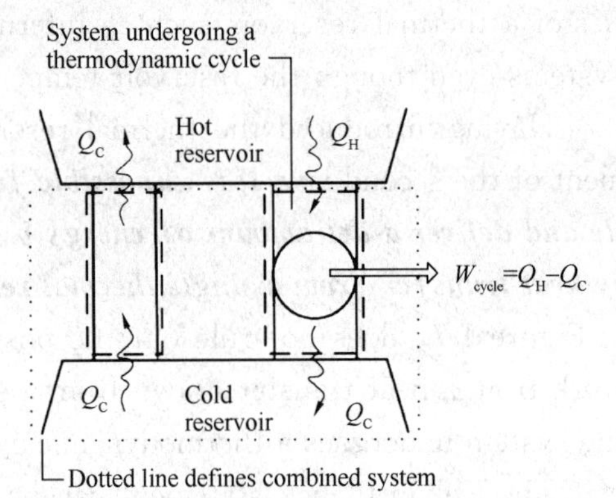

Figure 1-8 Illustration used to demonstrate the equivalence of the Clausius and Kelvin-Planck statements of the second law

Unit 2 Heat Transfer

Heat naturally flows from a higher energy level to a lower energy level. In other words, heat travels from a warmer substance to a cooler substance. Therefore, heat energy flows downhill and the expression"heat rises" is incorrect, it's heated air that rises and cooled air falls.

When there is a temperature difference between two substances, heat transfer will occur. In fact, temperature difference is the driving force for heat transfer. The greater the temperature difference is, the greater the heat transfer is. The three modes of heat transfer are conduction, convection and radiation. All modes of heat transfer require the existence of a temperature difference, and all modes are from the high-temperature medium to a lower- temperature one.

2.1 Conduction

Heat transfer by *conduction* is heat energy traveling from one molecule to another. A heat exchanger, for example, a water coil, in a commercial HVACR (Heating, Ventilating and Air Conditioning, Refrigeration) system or home furnace uses conduction to transfer heat. Your hand touching a cold wall is an example of heat transfer by conduction. However, heat does not conduct at the same rate in all materials. For example, metals are good conductors, but copper, a very good conductor, conducts at a different rate from iron. Glass, wood and air on the other hand are poor conductors of heat. Very poor conductors are called insulators.

Conduction can take place in solids, liquids, or gases. In gases and liquids, conduction is due to the collisions and *diffusion* of the molecules during their random motion. In solids, it is due to the combination of *vibrations* of the molecules in a lattice and the energy transport by free electrons. A cold canned drink in a warm room, for example, eventually warms up to the room temperature as a result of heat transfer from the room to the drink through the aluminum can by conduction.

The *rate* of heat conduction though a medium depends on the *geometry* of the medium, its thickness, and the *material* of the medium, as well as the temperature difference across the medium. We know that wrapping a hot water tank with glass wool (an insulating material) reduces the rate of heat loss from the tank. The thicker the insulation, the smaller the heat loss. We also know that a hot water tank will lose heat at a higher rate when the temperature of the room housing the tank is lowered. Further, the larger the tank, the larger the surface area and thus the rate of heat loss.

Consider steady heat conduction through a large plane wall of thickness $\Delta x = L$ and area A, as shown in Figure 2-1. The temperature difference across the wall is $T = T_2 - T_1$. Experiments have shown that the rate of heat transfer Q through the wall is doubled when the temperature differencef ΔT across the wall or the area A normal to the direction of heat transfer is doubled, but is halved when the wall thickness L is doubled. Thus we conclude that *the rate of heat conclusion through a plane layer is proportional to the temperature difference across the layer and the heat transfer area, but is inversely proportional to the thickness of the layer.* That is,

$$\text{Rate of heat conduction} \propto \frac{(\text{Area})(\text{Temperature difference})}{\text{Thickness}}$$

or,

$$\dot{Q}_{\text{cond}} = kA\frac{T_1 - T_2}{\Delta x} = -kA\frac{\Delta T}{\Delta x}(\text{W}) \tag{2-1}$$

where the constant of proportionality k is the *thermal conductivity* of the material, which is *a measure of the ability of a material to conduct heat* as shown in Figure 2-2. In the limiting case ofo $\Delta x \to 0$ In the limiting case of a material to conduct heatrop

$$\dot{Q}_{\text{cond}} = -kA\frac{\mathrm{d}T}{\mathrm{d}x}\ (\text{W}) \tag{2-2}$$

which is called *Fourier's law of heat conduction* after J. Fourier, who expressed it first in his heat transfer text in 1822. Here $\mathrm{d}T / \mathrm{d}x$ is the *temperature gradient*, which is the slope of the temperature curve on a T-x diagram (the rate of change of T with x), at location x. The relation above indicates that the rate of heat conduction in a direction is proportional to the temperature gradient in that direction, Heat is conducted in the direction of decreasing temperature, and the temperature gradient becomes negative when temperature decreases with increasing x. The *negative sign* in Equation 2-2 ensures that heat transfer in the positive x direction is a positive quantity.

The heat transfer area$A$ is always *normal* to the direction of heat transfer. For heat loss through a 5-m-long, 3-m-high, and 25-cm-thick wall, for example, the heat transfer area is $A = 15\ \text{m}^2$. Note that the thickness of the wall has no effect on A as shown in Figure 2-3.

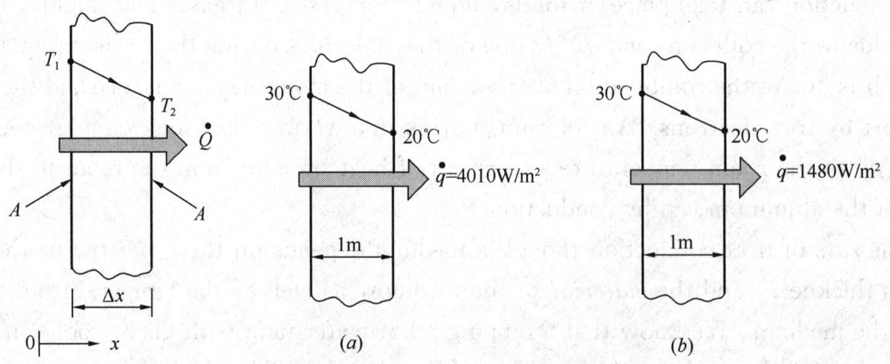

Figure 2-1 Heat conduction through a large plane wall.

Figure 2-2 The rate of heat conduction through a solid is directly proportional to its thermal conductivity.
(*a*) Copper (k=401 W/m·℃); (*b*) Silicon (k=148 W/m·℃)

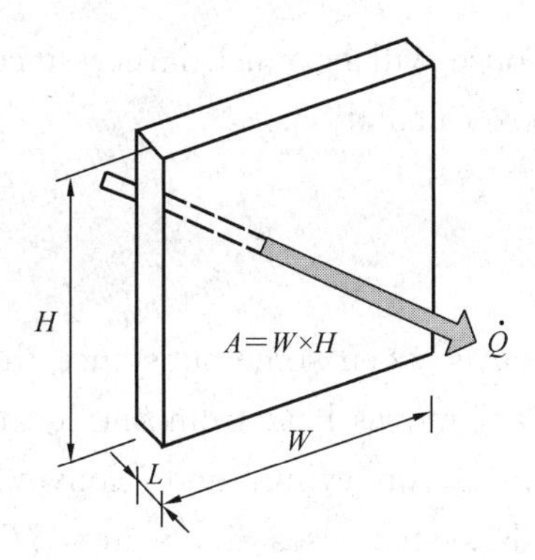

Figure 2-3 In heat conduction analysis, A represents the area *normal* to the direction of heat transfer

EXAMPLE 2-1 The Cost of Heat Loss through a Roof

The roof of an electrically heated home is 6 m long, 8 m wide, and 0.25 m thick, and is made of a flat layer of concrete whose thermal conductivity is $k = 0.8\mathrm{W/(m \cdot ℃)}$, as shown in Figure 2-4. The temperatures of the inner and the outer surfaces of the roof one night are measured to be 15℃ and 4℃, respectively, for a period of 10 hours. Determine (a) the rate of heat loss through the roof that night and (b) the cost of that heat loss to the home owner if the cost of electricity is \$0.08/kWh.

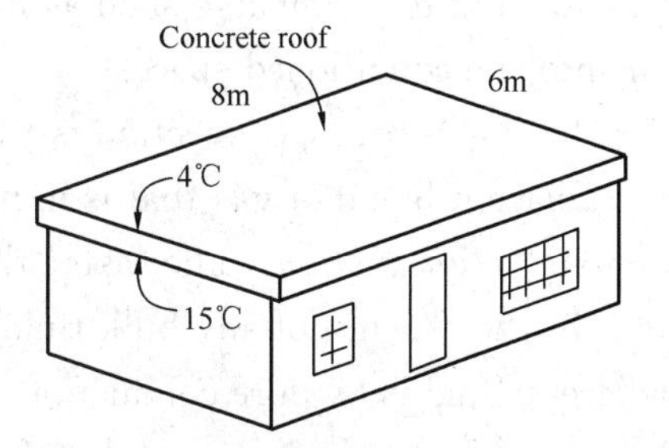

Figure 2-4 Schematic for Example 2-1.

SOLUTION The inner and outer surfaces of the flat concrete roof of an electrically heated home are maintained at specified temperatures during a night. The heat loss through the roof and its cost that night are to be determined.

Assumptions

1 Steady operating conditions exist during the entire night since the surface temperatures of the roof remain constant at the specified values.

2 Constant properties can be used for the roof.

Properties The thermal conductivity of the roof is given to be $k = 0.8\mathrm{W/m \cdot ℃}$.

Analysis (a) Noting that heat transfer through the roof is by conduction and the area of the roof is $A = 6\mathrm{m} \times 8\mathrm{m} = 48\mathrm{m}^2$ that heat t^2, the steady rate of heat transfer through the roof is determined to be

$$\dot{Q} = kA\frac{T_1 - T_2}{\Delta x} = (0.8\ \mathrm{W/m \cdot ℃})(48\ \mathrm{m}^2)\frac{(15-4)℃}{0.25\mathrm{m}} = 1690\mathrm{W} = 1.69\mathrm{kW}$$

(b) The amount of heat loss through the roof during a 10-hour period and its cost are determined from

$$Q = \dot{Q}\Delta t = (1.69\mathrm{kW})(10\mathrm{h}) = 16.9\mathrm{kWh}$$

$$\begin{aligned}\text{Cost} &= (\text{Amount of energy})(\text{Unit cost of energy})\\ &= (16.9\ \mathrm{kWh})(\$0.08/\mathrm{kWh}) = \$1.35\end{aligned}$$

total heating bill of the house will be much larger since the heat losses through the walls are not considered in these calculations.

2. 2 Convection

Heat transfer by *convection* is when some substance that is readily movable such as air, water, steam, or refrigerant moves heat from one location to another. Compare the words "convection" (the action of conveying) and "convey" (to take or carry from one place to another). An HVACR system uses air, water, steam and refrigerants in ducts and piping to convey heat energy to various parts of the system. As stated, when air is heated, it rises; this is heat transfer by "natural" convection. "Forced" convection is when a fan or pump is used to convey heat in fluids such as air and water. For example, many large buildings have a central heating plant where water is heated and pumped throughout the building to terminals, such as heating coils (aka heat exchangers). Fans move heated air into the conditioned space.

In engineering, convection is the mode of energy transfer between a solid surface and the adjacent liquid or gas that is in motion, and it involves the combined effects of *conduction* and *fluid motion*. The faster the fluid motion, the greater the convection heat transfer. In the absence of any bulk fluid motion, heat transfer between a solid surface and the adjacent fluid is by pure conduction. The presence of bulk motion of the fluid enhances the heat transfer between the solid surface and the fluid, but it also complicates the determination of heat transfer rates.

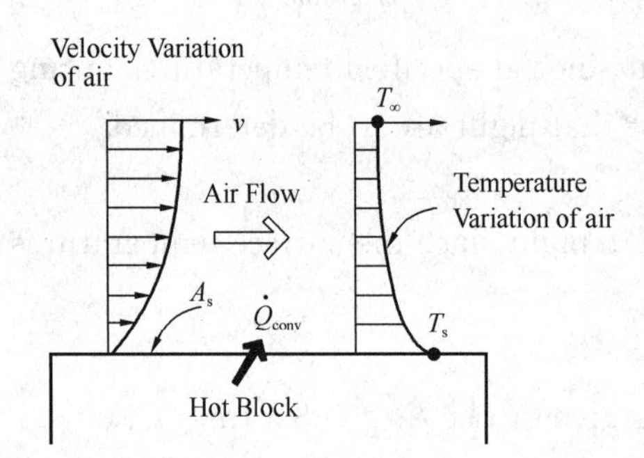

Figure 2-5 Heat transfer from a hot surface to air by convection

Consider the cooling of a hot block by blowing cool air over its top surface as shown in Figure 2-5. Energy is first transferred to the air layer adjacent to the block by conduction. This energy is then carried away from the surface by convection, that is, by the combined effects of conduction within the air that is due to random motion of air molecules and the bulk or macroscopic motion of the air that removes the heated air near the surface and replaces it by the cooler air.

As stated above, convection is called forced convection if the fluid is forced to flow over the surface by external means such as a fan, pump, or the wind. In contrast, convection is called natural (or free) *convection* if the fluid motion is caused by buoyancy forces that are induced by density differences due to the variation of temperature in the fluid as shown in Figure 2-6. For example, in the absence of a fan, heat transfer from the surface of the hot block in Figure 2-5 will be by natural convection since any motion in the air in this case will be due to the rise of the warmer (and thus lighter) air near the surface and the fall of the cooler (and thus heavier) air to fill its place. Heat transfer between the

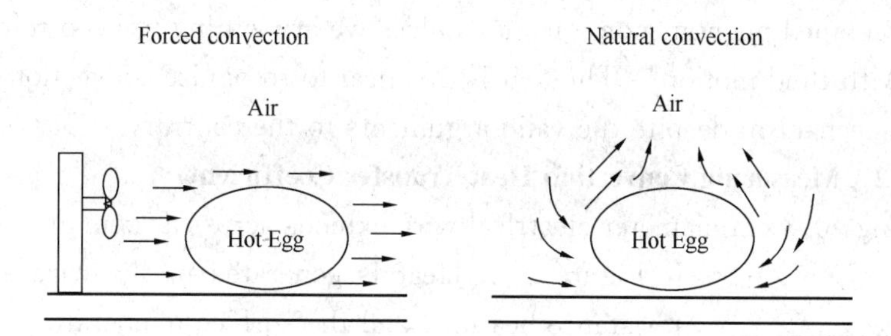

Figure 2-6 The cooling of a boiled egg by forced and natural convection

block and the surrounding air will be by conduction if the temperature difference between the air and the block is not large enough to overcome the resistance of air to movement and thus to initiate natural convection currents.

Heat transfer processes that involve *change of phase* of a fluid are also considered to by convection because of the fluid motion induced during the process, such as the rise of the vapor bubbles during boiling or the fall of the liquid droplets during condensation.

Despite the complexity of convection, the rate of convection heat transfer is observed to be proportional to the temperature difference, and is conveniently expressed by Newton's law of cooling as

$$\dot{Q}_{conv} = hA_s(T_s - T_\infty) \quad (W) \tag{2-3}$$

where h —— convection *heat transfer coefficient*, W/(m^2 • ℃) (Btu/(h • ft^2 • °F)

A_s—— surface area through which convection heat transfer takes place

T_s—— surface temperature

T_∞——temperature of the fluid sufficiently far from the surface

Note that at the surface, the fluid temperature equals the surface temperature of the solid.

The convection heat transfer coefficient h is not a property of the fluid. It is an experimentally determined parameter whose value depends on all the variables influencing convection such as the surface geometry, the nature of fluid motion, the properties of the fluid, and the bulk fluid velocity. Typical values of h are given in Table 2-1.

Typical values of convection heat transfer coefficient **Table 2-1**

Type of convection	h, W/m^2 • ℃ *
Free convection of gases	2—25
Free convection of liquids	10—1,000
Forced convection of gases	25—250
Forced convection of liquids	50—20,000
Boiling and condensation	2500—100,000

* Multiply by 0.176 to convert to Btu/h • ft^2 • °F.

Some people do not consider convection to be a fundamental mechanism of heat transfer since it is essentially heat conduction in the presence of fluid motion. But we still need to give this combined phenomenon a name, unless we are willing to keep referring to it as "conduction with fluid motion". Thus, it is practical to recognize convection as a separate heat transfer mechanism despite the valid arguments to the contrary.

EXAMPLE 2-2 Measuring Convection Heat Transfer Coefficient

A 2-m-long, 0. 3-cm-diameter electrical wire extends across a room at 15℃, as shown in Figure 2-7. Heat is generated in the wire as a result of resistance heating, and the surface temperature of the wire is measured to be 152℃ in steady operation. Also, the voltage drop and electric current through the wire are measured to be 60 V and 1. 5 A, respectively. Disregarding any heat transfer by radiation, determine the convection heat transfer coefficient for heat transfer between the outer surface of the wire and the air in the room.

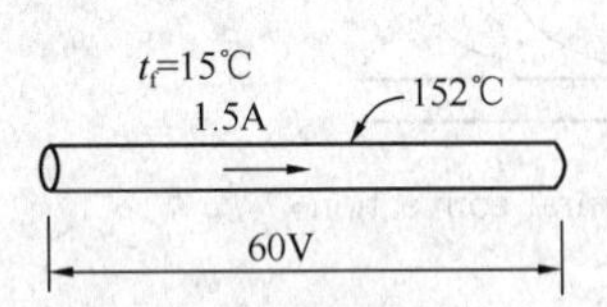

Figure 2-7 Schematic for Example 2-2

SOLUTION The convection heat transfer coefficient for heat transfer from an electrically heated wire to air is to be determined by measuring temperatures when steady operating conditions are reached and the electric power consumed.

Assumptions

1. Steady operating conditions exist since the temperature reading do not change with time.
2. Radiation heat transfer is negligible.

Analysis When steady operating conditions are reached, the rate of heat loss from the wire will equal the rate of heat generation in the wire as a result of resistance heating. That is

$$\dot{Q} = \dot{E}_{generated} = VI = (60V)(1.5A) = 90W$$

The surface area of wire is

$$A_s = \pi DL = \pi(0.003m)(2m) = 0.01885m^2$$

Newton's law of cooling for convection heat transfer is expressed as

$$\dot{Q}_{conv} = hA_s(T_s - T_\infty)$$

Disregarding any heat transfer by radiation and thus assuming all the heat loss from the wire to occur by convection, the convection heat transfer coefficient is determined to be

$$h = \frac{\dot{Q}_{conv}}{A_s(T_s - T_\infty)} = \frac{90W}{(0.01885\ m^2)(152-15)℃} = 34.9W/(m^2 \cdot ℃)$$

Discussion Not that the simple setup described above can be used to determine the average heat transfer coefficients from a variety of surfaces in air. Also, heat transfer by radiation can be eliminated by keeping the surrounding surfaces at the temperature of the wire.

2. 3 Radiation

Heat transferred by *radiation* travels through space without heating the space. Radiation (aka radiant heat) does not transfer the actual temperature value. A portable electric space heater is example of heat transfer by radiation. As the electric heater coil glows red-hot it radiates heat into the room. The space heater does not heat the air (the space) instead it warms the solid objects that the radiant heat encounters. However, radiant heat diminishes by the distance traveled. An example of diminishing heat is the sun. The earth does not experience the total heat of the sun—approximately 27,000,000°F at the sun's core but only about 10,000F where the sun's radiation is detected as sunlight to the earth—because the sun is some 93 million miles from the earth.

Radiation is the energy emitted by matter in the form of electromagnetic waves (*or* photons) as a result of the changes in the electronic configurations of the atoms or molecules. Unlike conduction and convection, the transfer of energy by radiation does not require the presence of an intervening medium. In fact, energy transfer by radiation is fastest (at the speed of light) and it suffers no attenuation in a vacuum. This is how the energy of the sun reaches the earth as mentioned above.

In heat transfer studies we are interested in *thermal radiation*, which is the form of radiation emitted by bodies because of their temperature. It differs from other forms of electromagnetic radiation such as x-rays, gamma rays, microwaves, radio waves, and television waves that are not related to temperature. All bodies at a temperature above absolute zero emit thermal radiation.

Radiation is a volumetric phenomenon, and all solids, liquids, and gases emit, absorb, or transmit radiation to varying degrees. However, radiation is usually considered to be a surface phenomenon for solids that are opaque to thermal radiation such as metals, wood, and rocks since the radiation emitted by the interior regions of such material can never reach the surface, and the radiation incident on such bodies is usually absorbed within a few microns from the surface.

The maximum rate of radiation that can be emitted from a surface at an absolute temperature T_s (in K or R) is given by the Stefan-Boltzmann law as

$$\dot{Q}_{emit,max} = \sigma A_s T_s^4 \quad (W) \tag{2-4}$$

where σ equals 5.67×10^{-8} W/($m^2 \cdot K^4$) or 0.1714 Btu/($h \cdot ft^2 \cdot R^4$) is the Stefan-Boltzmann constant. The idealized surface that emits radiation at this maximum rate is called a *blackbody*, and the radiation emitted by a blackbody is called blackbody radiation as shown in Figure 2-8. The radiation emitted by all real surfaces is less than the radiation emitted by a blackbody at the same temperature, and is expressed as

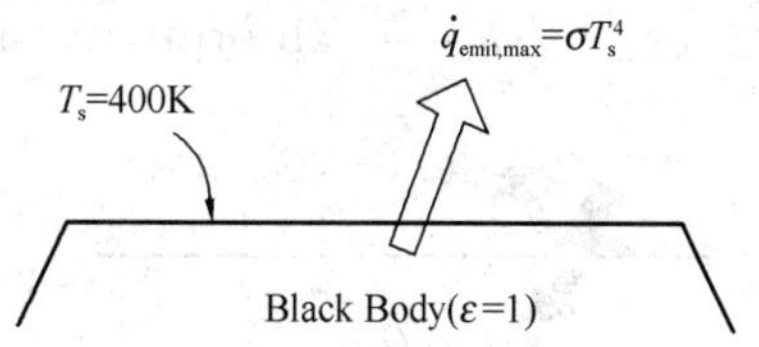

$$\dot{Q}_{emit} = \varepsilon \sigma A T_s^4 \quad (W) \tag{2-5}$$

Figure 2-8 Blackbody radiation represents the *maximum amount of radiation that can be emitted from a surface at a specified temperature*

where emissivity of the surface. The property emissivity, whose value is in the range $0 \leqslant \varepsilon \leqslant 1$, the ε is a measure of how closely a surface approximates a blackbody for whichemitted byemissivities of some surfaces are given in Table 2-2.

Emissivities of some materials at 300K **Table 2-2**

Material	Emissivity
Aluminum foil	0. 07
Anodized aluminum	0. 82
Polished copper	0. 03
Polished gold	0. 03
Polished silver	0. 02
Polished stainless steel	0. 17
Black paint	0. 98
White paint	0. 90
White paper	0. 92-0. 97
Asphalt pavement	0. 85-0. 93
Red brick	0. 93-0. 96
Human skin	0. 95
Wood	0. 82-0. 92
Soil	0. 93-0. 96
Water	0. 96
Vegetation	0. 92-0. 96

Another important radiation property of a surface is its absorptivity bsorptivityportant radiation property of a surface is its face at a specified temperaturebody at the same temperature, and is expressed ash material can never reach the surfacethe entire radiation incident on it. That is, a blackbody is a perfect absorber ($\alpha = 1$) as it is a perfect emitter.

In general, both ε and α of a surface depend on the temperature and the wavelength of the radiation. *Kirchhoff's law* of radiation states that the emissivity and the absorptivity of a surface at a given temperature and wavelength are equal. In many practical applications, the surface temperature and the temperature of the source of incident radiation are of the same order of magnitude, and the average absorptivity of a surface is taken to be equal to its average emissivity. The rate at which a surface absorbs radiation is determined from as shown in Figure 2-9.

$$\dot{Q}_{absorbed} = \alpha \dot{Q}_{incident} \quad (W) \tag{2-6}$$

where $\dot{Q}_{incident}$——rate at which radiation is incident on the surface, W

α —— absorptivity of the surface

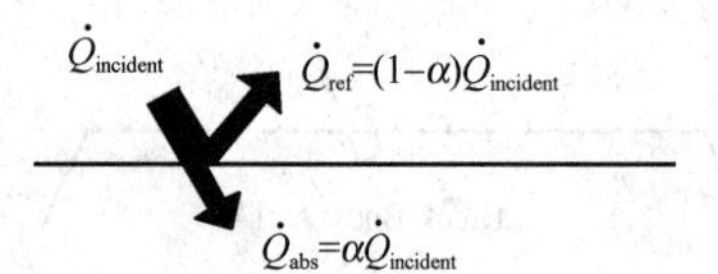

Figure 2-9 The absorption of radiation incident on an opaque surface of absorptivity α.

For opaque (nontransparent) surfaces, the portion of incident radiation not absorbed by the surface is reflected back.

The difference between the rates of radiation emitted by the surface and the radiation absorbed is the *net* radiation heat transfer. If the rate of radiation absorption is greater than the rate of radiation emission, the surface is said to be gaining energy by radiation. Otherwise, the surface is said to be losing energy by radiation. In general, the determination of the net rate of heat transfer by radiation between two surfaces is a complicated matter since it depends on the properties of the surfaces, their orientation relative to each other, and the interaction of the medium between the surfaces with radiation.

When a surface of emissivity and surface area A_s at an absolute temperature T_s is completely enclosed by a much larger (or black) surface at absolute temperature T_{surr} separated by a gas (such as air) that does not intervene with radiation, the net rate of radiation heat transfer between these two surfaces is given by (Figure 2-10)

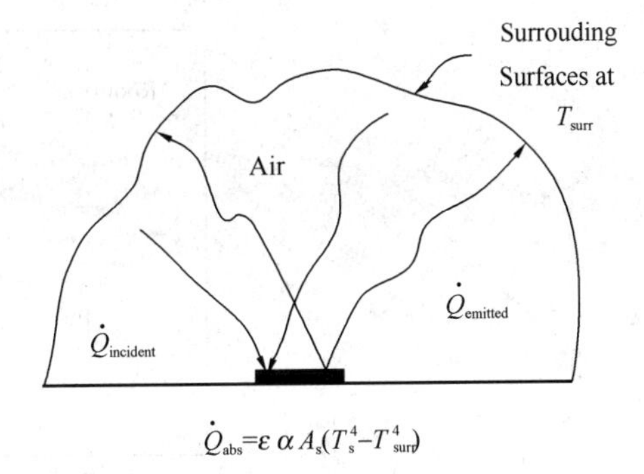

Figure 2-10 Radiation heat transfer between a surface and the surfaces surrounding it

$$\dot{Q}_{rad} = \varepsilon\sigma A_s(T_s^4 - T_{surr}^4) \quad (W) \tag{2-7}$$

In this special case, the emissivity and the surface area of the surrounding surface do not have any effect on the net radiation heat transfer.

Radiation heat transfer to or from a surface surrounded by a gas such as air occurs parallel to conduction (or convection, if there is bulk gas motion) between the surface and the gas. Thus the total heat transfer is determined by adding the contributions of both heat transfer mechanisms. For simplicity and convenience, this is often done by defining a combine heat transfer coefficient $h_{combined}$ that includes the effects of both convection and radiation. Then the *total* heat transfer rate to or from a surface by convection and radiation is expressed as

$$\dot{Q}_{total} = h_{combined} A_s(T_s - T_\infty) \quad (W) \tag{2-8}$$

Note that the combined heat transfer coefficient is essentially a convection heat transfer coefficient modified to include the effects of radiation.

Radiation is usually significant relative to conduction or natural convection, but negligible relative to forced convection. Thus radiation in forced convection applications is usually disregarded, especially when the surfaces involved have low emissivities and low to moderate temperatures.

EXAMPLE 2-3 Radiation Effect on Thermal Comfort

It is common experience to feel"chilly" in winter and "warm" in summer in our homes even when the thermostat setting is kept the same. This is due to the so called "radiation effect" resulting from radiation heat exchange between our bodies and the surrounding surfaces of the walls and the ceiling.

Consider a person standing in a room maintained at 22℃ at all times. The inner surfaces of the walls, floors, and the ceiling of the house are observed to be at an average temperature of 10℃ in winter and 25℃ in summer. Determine the rate of radiation heat transfer between this person and the surrounding surfaces if the exposed surface area and the average outer surface temperature of the person are 1.4m^2 and 30℃, respectively (Figure 2-11).

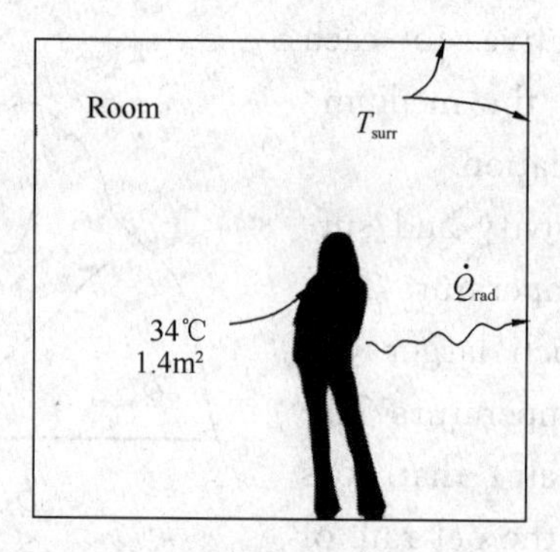

Figure 2-11 Schematic for Example 2-3

SOLUTION The rate of radiation heat transfer between a person and the surrounding surfaces at specified temperatures are to be determined in summer and winter.

Assumptions

1. Steady operating conditions exist.
2. Heat transfer by convection is not considered.
3. The person is completely surrounded by the interior surfaces of the room.
4. The surrounding surfaces are at a uniform temperature.

Properties The emissivity of a person is $\varepsilon=0.95$ (Table 2-2).

Analysis The net rates of radiation heat transfer from the body to the surrounding walls, ceiling, and floor in winter and summer are

$$\begin{aligned}\dot{Q}_{rad,winter} &= \varepsilon\sigma A_s(T_s^4 - T_{surr,winter}^4)\\ &= (0.95)(5.67\times10^{-8}\ \mathrm{W/(m^2\cdot K^4)})(1.4\ \mathrm{m^2})\\ &\quad\times[(30+273)^4-(10+273)^4]\mathrm{K^4}\\ &= 152\mathrm{W}\end{aligned}$$

and

$$\begin{aligned}\dot{Q}_{rad,summer} &= \varepsilon\sigma A_s(T_s^4 - T_{surr,summer}^4)\\ &= (0.95)(5.67\times10^{-8}\ \mathrm{W/(m^2\cdot K^4)})(1.4\ \mathrm{m^2})\\ &\quad\times[(30+273)^4-(25+273)^4]\mathrm{K^4}\\ &= 40.9\ \mathrm{W}\end{aligned}$$

Discussion Note that we must use absolute temperatures in radiation calculations. Also note that the rate of heat loss from the person by radiation is almost four times as large in winter than it is in summer, which explains the "chill" we feel in winter even if the thermostat setting is kept the same.

Unit 3　Fluid Mechanics

Fluid mechanics is the study of fluids either in motion (fluid *dynamics*) or at rest (fluid *statics*). Both gases and liquids are classified as fluids, and the number of fluid engineering applications is enormous: breathing, blood flow, swimming, pumps, fans, turbines, airplanes, ships, rivers, windmills, pipes, missiles, icebergs, engines, filters, jets, and sprinklers, to name a few. When you think about it, almost everything on this planet either is a fluid or moves within or near a fluid.

The essence of the subject of fluid flow is a judicious compromise between theory and experiment. Since fluid flow is a branch of mechanics, it satisfies a set of well-documented basic laws, and thus a great deal of theoretical treatment is available. However, the theory is often frustrating because it applies mainly to idealized situations, which may be invalid in practical problems. The two chief obstacles to a workable theory are geometry and viscosity. The basic equations of fluid motion are too difficult to enable the analyst to attack arbitrary geometric configurations. Thus most textbooks concentrate on flat plates, circular pipes, and other easy geometries. It is possible to apply numerical computer techniques to complex geometries, and specialized textbooks are now available to explain the new *computational fluid dynamics* (*CFD*) approximations and methods. The textbook by Frank M. White will present many theoretical results while keeping their limitations in mind.

The second obstacle to a workable theory is the action of *viscosity*, which can be neglected only in certain idealized flows. First, viscosity increases the difficulty of the basic equations, although the boundary-layer approximation found by Ludwig Prandtl in 1904 has greatly simplified viscous-flow analyses. Second, viscosity has a destabilizing effect on all fluids, giving rise, at frustratingly small velocities, to a disorderly, random phenomenon called turbulence. The theory of turbulent flow is crude and heavily backed up by experiment, yet it can be quite serviceable as an engineering estimate. The textbook by Frank M. White only introduces the standard experimental correlations for turbulent time-mean flow. Meanwhile, there are advanced texts on both time-mean turbulence and turbulence modeling and on the newer, computer-intensive direct numerical simulation (DNS) of fluctuating turbulence.

Thus there is theory available for fluid flow problems, but in all cases it should be backed up by experiment. Often the experimental data provide the main source of information about specific flows, such as the drag and lift of immersed bodies. Fortunately, fluid mechanics is a highly visual subject, with good instrumentation, and the use of dimensional analysis and modeling concepts is widespread. Thus experimentation provides a natural and easy complement to the theory. You should keep in mind that theory and experiment should go hand in hand in all studies of fluid mechanics.

3.1 Introduction

History and Scope of Fluid Mechanics

Like most scientific disciplines, fluid mechanics has a history of erratically occurring early achievements, then an intermediate era of steady fundamental discoveries in the eighteenth and nineteenth centuries, leading to the twenty-first-century era of "modern practice," as we self-centeredly term our limited but up-to-date knowledge. Ancient civilizations had enough knowledge to solve certain flow problems. Sailing ships with oars and irrigation systems were both known in prehistoric times. The Greeks produced quantitative information. Archimedes and Hero of Alexandria both postulated the parallelogram law for addition of vectors in the third century B. C. Archimedes (285 - 212 B. C.) formulated the laws of buoyancy and applied them to floating and submerged bodies, actually deriving a form of the differential calculus as part of the analysis. The Romans built extensive aqueduct systems in the fourth century B. C. but left no records showing any quantitative knowledge of design principles. From the birth of Christ to the Renaissance there was a steady improvement in the design of such flow systems as ships and canals and water conduits but no recorded evidence of fundamental improvements in flow analysis. Then Leonardo da Vinci (1452 - 1519) stated the equation of conservation of mass in one-dimensional steady flow. Leonardo was an excellent experimentalist, and his notes contain accurate descriptions of waves, jets, hydraulic jumps, eddy formation, and both low-drag (streamlined) and high-drag (parachute) designs. A Frenchman, Edme Mariotte (1620 - 1684), built the first wind tunnel and tested models in it.

Figure 3-1 Leonhard Euler (1707 ~ 1783) (Courtesy of the School of Mathematics and Statistics, University of St Andrew, Scotland)

Problems involving the momentum of fluids could finally be analyzed after Isaac Newton (1642 - 1727) postulated his laws of motion and the law of viscosity of the linear fluids now called newtonian. The theory first yielded to the assumption of a "perfect" or frictionless fluid, and the eighteenth-century mathematicians (Daniel Bernoulli, Leonhard Euler, Jean d'Alembert, Joseph-Louis Lagrange, and Pierre-Simon Laplace) produced many beautiful solutions of frictionless-flow problems. Euler who published over 800 books and papers, Figure 3-1, developed both the differential equations of motion and their integrated form, now called the Bernoulli equation. Leonhard Euler was the greatest mathematician of the eighteenth century and used Newton's calculus to develop and solve the equations of motion of inviscid flow. D'Alembert used them to show his famous paradox: that a body immersed in a frictionless fluid has zero drag. These beautiful results amounted to overkill, since perfect-fluid assumptions have very limited application in practice and most engineering flows are dominated by the effects of viscosity. Engineers began to reject what they regarded as a totally unrealistic theory and developed the science of hydraulics, relying almost entirely on experiment. Such experimentalists as Chézy, Pitot, Borda, Weber, Francis, Hagen, Poiseuille, Darcy, Manning, Bazin, and Weisbach produced data on a variety of flows such as open channels, ship resistance, pipe flows, waves, and turbines. All the data were too often used in raw form without regard to the fundamental physics of flow.

Figure 3-2 Ludwig Prandtl (1875-1953)(Aufnahme von Fr. Struckmeyer, Gottingen, courtesy AIP Emilio Segre Visual Archives, Lande Collection)

At the end of the nineteenth century, unification between experimental hydraulics and theoretical hydrodynamics finally began. William Froude (1810-1879) and his son Robert (1846-1924) developed laws of model testing; Lord Rayleigh (1842-1919) proposed the technique of dimensional analysis; and Osborne Reynolds (1842-1912) published the classic pipe experiment in 1883, which showed the importance of the dimensionless Reynolds number named after him. Meanwhile, viscous-flow theory was available but unexploited, since Navier (1785-1836) and Stokes (1819-1903) had successfully added Newtonian viscous terms to the equations of motion. The resulting Navier-Stokes equations were too difficult to analyze for arbitrary flows. Then, in 1904, a German engineer, Ludwig Prandtl (1875-1953), who often been called the "father of modern fluid mechanics", Figure 3-2, published perhaps the most important paper ever written on fluid mechanics. He developed boundary layer theory and many other innovative analyses. He and his students were pioneers in flow visualization techniques. Prandtl pointed out that fluid flows with small viscosity, such as water flows and airflows, can be divided into a thin viscous layer, or boundary layer, near solid surfaces and interfaces, patched onto a nearly inviscid outer layer, where the Euler and Bernoulli equations apply. Boundary-layer theory has proved to be a very important tool in modern flow analysis. The twentiethcentury foundations for the present state of the art in fluid mechanics were laid in a series of broad-based experiments and theories by Prandtl and his two chief friendly competitors, Theodore von Kármán (1881-1963) and Sir Geoffrey I. Taylor (1886-1975). Many of the results sketched here from a historical point of view will, of course, be discussed in the textbook by Frank M. White.

The second half of the twentieth century introduced a new tool: Computational Fluid Dynamics (CFD). The earliest paper on the subject known to this writer was by A. Thom in 1933, a laborious, but accurate, hand calculation of flow past a cylinder at low Reynolds numbers. Commercial digital computers became available in the 1950s, and personal computers in the 1970s, bringing CFD into adulthood. A legendary first textbook was by Patankar. Presently, with increases in computer speed and memory, almost any laminar flow can be modeled accurately. Turbulent flow is still calculated with empirical models, but Direct Numerical Simulation is possible for low Reynolds numbers. Another five orders of magnitude in computer speed are needed before general turbulent flows can be calculated. That may not be possible, due to size limits of nano- and pico-elements. But, if general DNS develops, Gad-el-Hak raises the prospect of a shocking future: all of fluid mechanics reduced to a black box, with no real need for teachers, researchers, writers, or fluids engineers.

Since the earth is 75 percent covered with water and 100 percent covered with air, the scope of fluid mechanics is vast and touches nearly every human endeavor. The sciences of meteorology, physical oceanography, and hydrology are concerned with naturally occurring fluid flows, as are medical studies of breathing and blood circulation. All transportation problems involve fluid motion, with well-developed specialties in aerodynamics of aircraft and rockets and in naval hydrodynamics of ships and submarines. Almost all our electric energy is developed either from water flow or from steam flow through turbine generators. All combustion problems involve fluid motion as do the more classic problems of irrigation, flood control, water supply, sewage disposal, projectile motion, and oil and gas pipelines. The aim of this book is to present enough fundamental concepts and practical applications in fluid mechanics to prepare you to move smoothly into any of these specialized fields of the science of flow, and then be prepared to move out again as new technologies develop.

Problem-Solving Techniques

Fluid flow analysis is packed with problems to be solved. The present text has more than 1700 problem assignments. Solving a large number of these is a key to learning the subject. One must deal with equations, data, tables, assumptions, unit systems, and solution schemes. The degree of difficulty will vary, and we urge you to sample the whole spectrum of assignments, with or without the Answers in the Appendix.

Here are the recommended steps for problem solution:

1. Read the problem and restate it with your summary of the results desired.

2. From tables or charts, gather the needed property data: density, viscosity, etc.

3. Make sure you understand what is asked. Students are apt to answer the wrong question, for example, pressure instead of pressure gradient, lift force instead of drag force, or mass flow instead of volume flow. Read the problem carefully.

4. Make a detailed, labeled sketch of the system or control volume needed.

5. Think carefully and list your assumptions. You must decide if the flow is steady or unsteady, compressible or incompressible, viscous orinviscid, and whether a control volume or partial differential equations are needed.

6. Find an algebraic solution if possible. Then, if a numerical value is needed, use ei-

ther the SI or BG unit systems, to be reviewed in Section 1.6.

7. Report your solution, labeled, with the proper units and the proper number of significant figures (usually two or three) that the data uncertainty allows. We shall follow these steps, where appropriate, in our example problems.

The Concept of a Fluid

From the point of view of fluid mechanics, all matter consists of only two states, fluid and solid. The difference between the two is perfectly obvious to the layperson, and it is an interesting exercise to ask a layperson to put this difference into words. The technical distinction lies with thereaction of the two to an applied shear or tangential stress. A solid can resist a shear stress by a static deflection; *a fluid cannot*. Any shear stress applied to a fluid, no matter how small, will result in motion of that fluid. The fluid moves and deforms continuously as long as the shear stress is applied. As a corollary, we can say that a fluid at rest must be in a state of zero shear stress, a state often called the hydrostatic stress condition in structural analysis. In this condition, Mohr's circle for stress reduces to a point, and there is no shear stress on any plane cut through the element under stress.

Given this definition of a fluid, every layperson also knows that there are two classes of fluids, *liquids* and *gases*. Again the distinction is a technical one concerning the effect of cohesive forces. A liquid, being composed of relatively close-packed molecules with strong cohesive forces, tends to retain its volume and will form a free surface in a gravitational field if unconfined from above. Free-surface flows are dominated by gravitational effects. Since gas molecules are widely spaced with negligible cohesive forces, a gas is free to expand until it encounters confining walls. A gas has no definite volume, and when left to itself without confinement, a gas forms an atmosphere that is essentially hydrostatic. Gases cannot form a free surface, and thus gas flows are rarely concerned with gravitational effects other than buoyancy.

In the previous discussion, clear decisions could be made about solids, liquids, and gases. Most engineering fluid mechanics problems deal with these clear cases, that is, the common liquids, such as water, oil, mercury, gasoline, and alcohol, and the common gases, such as air, helium, hydrogen, and steam, in their common temperature and pressure ranges. There are many borderline cases, however, of which you should be aware. Some apparently "solid" substances such as asphalt and lead resist shear stress for short periods but actually deform slowly and exhibit definite fluid behavior over long periods. Other substances, notably colloid and slurry mixtures, resist small shear stresses but "yield" at large stress and begin to flow as fluids do. Specialized textbooks are devoted to this study of more general deformation and flow, a field called rheology. Also, liquids and gases can coexist in two-phase mixtures, such as steam-water mixtures or water with entrapped air bubbles. Specialized textbooks present the analysis of such multiphase flows. Finally, in some situations the distinction between a liquid and a gas blurs. This is the case at temperatures and pressures above the so-called critical point of a substance, where only a single phase exists, primarily resembling a gas. As pressure increases far above the critical point, the gaslike substance becomes so dense that there is some resemblance to a liquid and the usual thermodynamic approximations like the perfect-gas law become inaccurate. The critical temperature and pressure of water are $T_c = 647$ K and $p_c = 219$ atm (atmosphere, and one atmosphere equals 2116 $lbf/ft^2 = 101,300$ Pa) so that typical problems involving water and steam are below the critical point. Air, being a mixture of gases, has no distinct critical point, but its principal component, nitrogen, has $T_c = 126$K and $p_c = 34$ atm. Thus typical problems involving air are in the range of high temperature and low pressure where air is distinctly and definitely a gas.

3.2 Differential Relations for Fluid Flow

In analyzing fluid motion, we might take one of two paths: (1) seeking an estimate of gross effects (mass flow, induced force, energy change) over a finite region or control volume or (2) seeking the point-by-point details of a flow pattern by analyzing an infinitesimal region of the flow.

This section treats the second in our trio of techniques for analyzing fluid motion: small-scale, or differential, analysis. That is, we apply our four basic conservation laws to an infinitesimally small control volume or, alternately, to an infinitesimal fluid system. In either case the results yield the basic differential equations of fluid motion. Appropriate boundary conditions are also developed. In their most basic form, these differential equations of motion are quite difficult to solve, and very little is known about their general mathematical properties. However, certain things can be done that have great educational value. First, the equations (even if unsolved) reveal the basic dimensionless parameters that govern fluid motion. Second, a great number of useful solutions can be found if one makes two simplifying assumptions: (1) steady flow and (2) incompressible flow. A third and rather drastic simplification, frictionless flow, makes our old friend the Bernoulli equation valid and yields a wide variety of idealized, or perfect-fluid, possible solutions. We must be careful to ascertain whether such solutions are in fact realistic when compared with actual fluid motion. Finally, even the difficult general differential equations now yield to the approximating technique known as computational fluid dynamics (CFD) whereby the derivatives are simulated by algebraic relations between a finite number of grid points in the flow field, which are then solved on a computer. Reference 1 is an example of a textbook devoted entirely to numerical analysis of fluid motion.

The Acceleration Field of a Fluid

The cartesian vector form of a velocity field that varies in space and time is given by

$$\vec{V}(\vec{r},t) = \vec{i}\,u(x,y,z,t) + \vec{j}\,v(x,y,z,t) + \vec{k}\,w(x,y,z,t) \tag{3-1}$$

This is the most important variable in fluid mechanics: Knowledge of the velocity vector field is nearly equivalent to solving a fluid flow problem. Our coordinates are fixed in space, and we observe the fluid as it passes by, as if we had scribed a set of coordinate lines on a glass window in a wind tunnel. This is theeulerian frame of reference, as opposed to the lagrangian frame, which follows the moving position of individual particles.

To write Newton's second law for an infinitesimal fluid system, we need to calculate the acceleration vector fielda of the flow. Thus we compute the total time derivative of the velocity vector:

$$\vec{a}=\frac{d\vec{V}}{dt}(\vec{r},t)=\vec{i}\frac{du}{dt}+\vec{j}\frac{dv}{dt}+\vec{k}\frac{dw}{dt} \tag{3-2}$$

Since each scalar component (u, v, w) is a function of the four variables (x, y, z, t), we use the chain rule to obtain each scalar time derivative. For example,

$$\frac{du(x,y,z,t)}{dt}=\frac{\partial u}{\partial t}+\frac{\partial u}{\partial x}\frac{dx}{dt}+\frac{\partial u}{\partial y}\frac{dy}{dt}+\frac{\partial u}{\partial z}\frac{dz}{dt} \tag{3-3}$$

But, by definition, dx/dt is the local velocity component u, and $dy/dt=v$, and $dz/dt=w$. The total time derivative of u may thus be written as follows, with exactly similar expressions for the time derivatives of v and w:

$$\begin{aligned}
a_x&=\frac{du}{dt}=\frac{\partial u}{\partial t}+u\frac{\partial u}{\partial x}+v\frac{\partial u}{\partial y}+w\frac{\partial u}{\partial z}=\frac{\partial u}{\partial t}+(\vec{V}\cdot\vec{\nabla})u\\
a_y&=\frac{dv}{dt}=\frac{\partial v}{\partial t}+u\frac{\partial v}{\partial x}+v\frac{\partial v}{\partial y}+w\frac{\partial v}{\partial z}=\frac{\partial v}{\partial t}+(\vec{V}\cdot\vec{\nabla})v\\
a_z&=\frac{dw}{dt}=\frac{\partial w}{\partial t}+u\frac{\partial w}{\partial x}+v\frac{\partial w}{\partial y}+w\frac{\partial w}{\partial z}=\frac{\partial w}{\partial t}+(\vec{V}\cdot\vec{\nabla})w
\end{aligned} \tag{3-4}$$

Summing these into a vector, we obtain the total acceleration:

$$\vec{a}=\frac{d\vec{V}}{dt}=\underbrace{\frac{\partial\vec{V}}{\partial t}}_{\text{Local}}+\underbrace{\left(u\frac{\partial\vec{V}}{\partial x}+v\frac{\partial\vec{V}}{\partial y}+w\frac{\partial\vec{V}}{\partial z}\right)}_{\text{Convective}}=\frac{\partial\vec{V}}{\partial t}+(\vec{V}\cdot\vec{\nabla})\vec{V} \tag{3-5}$$

The term $\partial\vec{V}/\partial t$ is called the local acceleration, which vanishes if the flow is steady, that is, independent of time. The three terms in parentheses are called the convective acceleration, which arises when the particle moves through regions of spatially varying velocity, as in a nozzle or diffuser. Flows that are nominally "steady" may have large accelerations due to the convective terms.

Note our use of the compact dot product involving$\vec{V}$and the gradient operator $\vec{\nabla}$:

$$u\frac{\partial}{\partial x}+v\frac{\partial}{\partial y}+w\frac{\partial}{\partial z}=\vec{V}\cdot\vec{\nabla}\qquad\text{where}\qquad\vec{\nabla}=\vec{i}\frac{\partial}{\partial x}+\vec{j}\frac{\partial}{\partial y}+\vec{k}\frac{\partial}{\partial z}$$

The total time derivative-sometimes called the substantial or material derivative-concept may be applied to any variable, such as the pressure:

$$\frac{dp}{dt}=\frac{\partial p}{\partial t}+\left(u\frac{\partial p}{\partial x}+v\frac{\partial p}{\partial y}+w\frac{\partial p}{\partial z}\right)=\frac{\partial p}{\partial t}+(\vec{V}\cdot\vec{\nabla})p \tag{3-6}$$

Wherever convective effects occur in the basic laws involving mass, momentum, or energy, the basic differential equations become nonlinear and are usually more complicated than flows that do not involve convective changes.

We emphasize that this total time derivative follows a particle of fixed identity, making it convenient for expressing laws of particle mechanics in theeulerian fluid field description. The operator d/dt is sometimes assigned a special symbol such as D/Dt as a further reminder that it contains four terms and follows a fixed particle.

As another reminder of the special nature of d/dt, some writers give it the name substantial or material derivative.

EXAMPLE 3-1 The Total Acceleration of a Particle

Given theeulerian velocity vector field

$$\vec{V} = 3t\,\vec{i} + xz\,\vec{j} + ty^2\,\vec{k}$$

find the total acceleration of a particle.

SOLUTION Carry out all the required derivatives with respect to (x, y, z, t), substitute into the total acceleration vector, Equation (3-5), and collect terms.

Assumptions

Given three known unsteady velocity components, $u = 3t$, $v = xz$, and $w = ty^2$.

Step 1: First work out the local acceleration $\partial\vec{V}/\partial t$:

$$\frac{\partial\vec{V}}{\partial t} = \vec{i}\frac{\partial u}{\partial t} + \vec{j}\frac{\partial v}{\partial t} + \vec{k}\frac{\partial w}{\partial t} = \vec{i}\frac{\partial}{\partial t}(3t) + \vec{j}\frac{\partial}{\partial t}(xz) + \vec{k}\frac{\partial}{\partial t}(ty^2)$$

$$= 3\vec{i} + 0\vec{j} + y^2\vec{k}$$

Step 2: In a similar manner, the convective acceleration terms, from Eq. (3-5), are

$$u\frac{\partial\vec{V}}{\partial x} = (3t)\frac{\partial}{\partial x}(3t\vec{i} + xz\vec{j} + ty^2\vec{k}) = (3t)(0\vec{i} + z\vec{j} + 0\vec{k}) = 3tz\vec{j}$$

$$v\frac{\partial\vec{V}}{\partial y} = (xz)\frac{\partial}{\partial y}(3t\vec{i} + xz\vec{j} + ty^2\vec{k}) = (xz)(0\vec{i} + 0\vec{j} + 2ty\vec{k}) = 2txyz\vec{k}$$

$$w\frac{\partial\vec{V}}{\partial z} = (ty^2)\frac{\partial}{\partial z}(3t\vec{i} + xz\vec{j} + ty^2\vec{k}) = (ty^2)(0\vec{i} + x\vec{j} + 0\vec{k}) = txy^2\vec{j}$$

Step 3: Combine all four terms above into the single "total" or "substantial" derivative:

$$\vec{a} = \frac{d\vec{V}}{dt} = \frac{\partial\vec{V}}{\partial t} + \left(u\frac{\partial\vec{V}}{\partial x} + v\frac{\partial\vec{V}}{\partial y} + w\frac{\partial\vec{V}}{\partial z}\right)$$

$$= (3\vec{i} + y^2\vec{k}) + 3tz\vec{j} + 2txyz\vec{k} + txy^2\vec{j}$$

$$= 3\vec{i} + (3tz + txy^2)\vec{j} + (y^2 + 2txyz)\vec{k}$$

Discussion Assuming that $\vec{V}$ is valid everywhere as given, this total acceleration vector $d\vec{V}/dt$ applies to all positions and times within the flow field.

3.3 Computational Fluid Dynamics

When potential flow involves complicated geometries or unusual stream conditions, the classical superposition scheme becomes less attractive. Conformal mapping of body shapes, by using the complex-variable technique, is no longer popular. Numerical analysis is the appropriate modern approach, and at least three different approaches are in use:

1. The finite element method (FEM)
2. The finite difference method (FDM)
3. a. Integral methods with distributed singularities
   b. The boundary element method

Methods 3a and 3b are closely related, having first been developed on an ad hoc basis by aerodynamicists in the 1960s and then generalized into a multipurpose applied mechanics technique in the 1970s. Methods 1 (or FEM) and 2 (or FDM), though strikingly different in concept, are comparable in scope, mesh size, and general accuracy. We concentrate here on the latter method for illustration purposes.

The Finite Element Method

The finite element method is applicable to all types of linear and nonlinear partial differential equations in physics and engineering. The computational domain is divided into small regions, usually triangular or quadrilateral. These regions are delineated with a finite number of nodes where the field variables—temperature, velocity, pressure, stream function, and so on—are to be calculated. The solution in each region is approximated by an algebraic combination of local nodal values. Then the approximate functions are integrated over the region, and their error is minimized, often by using a weighting function. This process yields a set of N algebraic equations for the N unknown nodal values. The nodal equations are solved simultaneously, by matrix inversion or iteration.

The Finite Difference Method

Although textbooks on numerical analysis apply finite difference techniques to many different problems, here we concentrate on potential flow. The idea of FDM is to approximate the partial derivatives in a physical equation by "differences" between nodal values spaced a finite distance apart-a sort of numerical calculus. The basic partial differential equation is thus replaced by a set of algebraic equations for the nodal values. For potential (inviscid) flow, these algebraic equations are linear, but they are generally nonlinear for viscous flows. The solution for nodal values is obtained by iteration or matrix inversion. Nodal spacings need not be equal.

Here we illustrate the two-dimensional Laplace equation, choosing for convenience the stream-function form

$$\frac{\partial^2 \psi}{\partial x^2} + \frac{\partial^2 \psi}{\partial y^2} = 0 \tag{3-7}$$

subject to known values of ψ along any body surface and known values of $\partial\psi/\partial x$ and $\partial\psi/\partial y$ in the free stream.

Our finite difference technique divides the flow field into equally spaced nodes, as shown in Figure 3-3. To economize on the use of parentheses or functional notation, subscripts i and j denote the position of an arbitrary, equally spaced node, and $\psi_{i,j}$ denotes the value of the stream

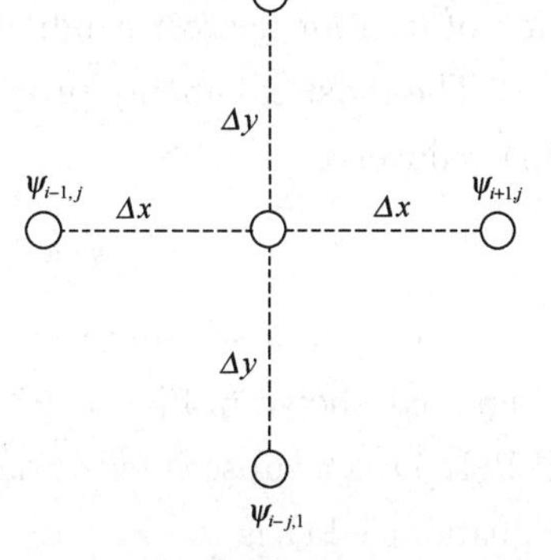

Figure 3-3 Definition sketch for a two-dimensional rectangular finite difference grid

function at that node:

$$\psi_{i,j} = \psi(x_0 + i\,\Delta x,\ y_0 + j\,\Delta y) \tag{3-8}$$

Thus $\psi_{i+1,j}$ is just to the right of $\psi_{i,j}$, and $\psi_{i,j+1}$ is just above.

An algebraic approximation for the derivative $\partial\psi/\partial x$ is

$$\frac{\partial \psi}{\partial x} \approx \frac{\psi(x+\Delta x, y) - \psi(x, y)}{\Delta x} \tag{3-9}$$

A similar approximation for the second derivative is

$$\frac{\partial^2 \psi}{\partial x^2} \approx \frac{1}{\Delta x}\left[\frac{\psi(x+\Delta x, y) - \psi(x, y)}{\Delta x} - \frac{\psi(x, y)\psi - (x-\Delta x, y)}{\Delta x}\right] \tag{3-10}$$

The subscript notation makes these expressions more compact:

$$\frac{\partial \psi}{\partial x} \approx \frac{1}{\Delta x}(\psi_{i+1,j} - \psi_{i,j})$$

$$\frac{\partial^2 \psi}{\partial x^2} \approx \frac{1}{\Delta x^2}(\psi_{i+1,j} - 2\psi_{i,j} + \psi_{i-1,j}) \tag{3-11}$$

These formulas are exact in the calculus limit as ex hese formin numerical analysis we keep Δx and Δy finite, hence the term finite differences.

In an exactly similar manner we can derive the equivalent differenceexpressionsfor the y direction:

$$\frac{\partial \psi}{\partial y} \approx \frac{1}{\Delta x}(\psi_{i,j+1} - \psi_{i,j})$$

$$\frac{\partial^2 \psi}{\partial x^2} \approx \frac{1}{\Delta x^2}(\psi_{i,j+1} - 2\psi_{i,j} + \psi_{i,j-1}) \tag{3-12}$$

The use of subscript notation allows these expressions to be programmed directly into a scientific computer language such as BASIC or FORTRAN.

When Equation(3-11) and Equation(3-12) are substituted into Laplace's Equation (3-7), the result is the algebraic formula

$$(1+\beta)\psi_{i,j} \approx \psi_{i-1,j} + \psi_{i+1,j} + \beta(\psi_{i,j-1} + \psi_{i,j+1}) \tag{3-13}$$

where $\beta = (ex\ /\ y)^2$ depends on the mesh size selected. This finite difference model of Laplace's equation states that every nodal stream-function value $\Psi_{i,j}$ is a linear combination of its four nearest neighbors.

The most commonly programmed case is a square mesh $\beta = 1$, for which Equation (3-13) reduces to

$$\psi_{i,j} \approx \frac{1}{4}(\psi_{i,j+1} + \psi_{i,j-1} + \psi_{i+1,j} + \psi_{i-1,j}) \tag{3-14}$$

Thus, for a square mesh, each nodal value equals the arithmetic average of the four neighbors shown in Figure 3-3. The formula is easily remembered and easily programmed. If P(I, J) is a subscripted variable stream function, the BASIC or FORTRAN statement of Eguation(3-14) is

$$P(I,J) \approx 0.25(P(I,J+1) + P(I,J-1) + P(I+1,J) + P(I-1,J)) \tag{3-15}$$

This is applied in iterative fashion sweeping over each of the internal nodes (I, J), with known values of P specified at each of the surrounding boundary nodes. Any initial guesses can be specified for the internal nodesP(I, J), and the iteration process will converge to the final algebraic solution in a finite number of sweeps. The numerical error, compared with the exact solution of Laplace's equation, is proportional to the square of the mesh size.

Convergence can be speeded up by the *successive overrelaxation* (SOR) method, discussed by Cebeci [5]. The modified SOR form of the iteration is

$$P(I,J) \approx P(I,J) + 0.25 \times A \times (P(I,J+1) + P(I,J-1) + P(I+1,J) + P(I-1,J) - 4P(I,J)) \quad (3\text{-}16)$$

The recommended value of the SOR convergence factor A is about 1.7. Note that the value A = 1.0 reduces Equation (3-16) to Equation (3-15).

Unit 4 Psychrometrics

Air properties and airflow are the heart of air conditioning. Technicians need to understand air properties and airflow to be able to troubleshoot and maintain an HVACR system. Factors that adversely affect system airflow will also have a negative impact on system efficiency and reliability. When people think about the properties of air and comfort, they think first about temperature. However, temperature is not the only property of air that is important to air conditioning. The properties of air include weight, volume, temperature, water content, and heat content. Humidity is nearly as important to human comfort as temperature. All the properties of air are important for the operation of air-conditioning equipment. The study of air and its properties is called *psychrometrics*.

Psychrometrics uses thermodynamic properties to analyze conditions and processes involving moist air. Formulas developed by Herrmann et al. (2009) may be used where greater precision is required. Herrmann et al. (2009), Hyland and Wexler (1983a, 1983b), and Nelson and Sauer (2002) developed formulas for thermodynamic properties of moist air and water modeled as real gases. However, perfect gas relations can be substituted in most air-conditioning problems. Kuehn et al. (1998) showed that errors are less than 0.7% in calculating humidity ratio, enthalpy, and specific volume of saturated air at standard atmospheric pressure for a temperature range of −60°F to 120°F. Furthermore, these errors decrease with decreasing pressure.

4.1 Air

Composition of dry and moist air

Atmospheric air contains many gaseous components as well as water vapor and miscellaneous contaminants (e.g., smoke, pollen, and gaseous pollutants not normally present in free air far from pollution sources).

Dry air is atmospheric air with all water vapor and contaminants removed. Its composition is relatively constant, but small variations in the amounts of individual components occur with time, geo-graphic location, and altitude. Harrison (1965) lists the approximate percentage composition of dry air by volume as: nitrogen, 78.084; oxygen, 20.9476; argon, 0.934; neon, 0.001818; helium, 0.000524; methane, 0.00015; sulfur dioxide, 0 to 0.0001; hydrogen, 0.00005 and minor components such as krypton, xenon, and ozone, 0.0002. Harrison (1965) and Hyland and Wexler (1983a) used a value 0.0314 (circa 1955) for carbon dioxide. Carbon dioxide reached 0.0379 in 2005, is currently increasing by 0. 00019 percent per year and is projected to reach 0. 0438 in 2036 (Gatley et al. 2008; Keeling and Whorf 2005a, 2005b). Increases in carbon dioxide are offset by decreases in oxygen; consequently, the oxygen percentage in 2036 is projected to be 20. 9352. Using the projected changes, the relative molecular mass for dry air for at least the first half of the 21st century is 28. 966, based on the carbon-12 scale. The gas constant for dry air using the current Mohr and Taylor (2005) values for the universal gas.

$$R_{da} = \frac{1545.349}{28.966} = 53.350\ (\text{ft} \cdot \text{lb}_f / \text{lb}_{da} \cdot {}^0R) \tag{4-1}$$

Moist air is a binary (two-component) mixture of dry air and water vapor. The amount of water vapor varies from zero (dry air) to a maximum that depends on temperature and pressure. Saturation is a state of neutral equilibrium between moist air and the condensed water phase (liquid or solid); unless otherwise stated, it assumes a flat interface surface between moist air and the condensed phase. Saturation conditions change when the interface radius is very small (e. g. , with ultrafine water droplets). The relative molecular mass of water is 18. 015268 on the carbon-12 scale. The gas constant for water vapor is

$$R_w = \frac{8314.472}{18.015268} = 461.524\ (\text{J/kg}_w \cdot \text{K}) \tag{4-2}$$

Equation of State of an Ideal Gas

The equation of state of an ideal gas indicates the relationship between its thermodynamic properties, or

$$pv = RT_R \tag{4-3}$$

where p ——pressure of gas, psf (Pa)

v ——specific volume of gas, ft^3/lb (m^3/kg)

R ——gas constant, ft · lbf / lbm · °R (J/kg · K)

T_R——absolute temperature of gas, °R (K)

Since $v = V/m$, then Eq. (4-3) becomes

$$pV = mRT_R \tag{4-4}$$

where V = total volume of gas, ft^3 (m^3)

m = mass of gas, lb (kg)

Using the relationship $m = nM$, and $R = R_o/M$, we can write Eq. (4-4) as

$$pV = nR_oT_R \tag{4-5}$$

where n ——number of moles, mol

M ——molecular weight

R_o——universal gas constant, ft · lbf / lbm · °R (J/mol · K)

Equation of State of a Real Gas

A modified form of the equation of state for a real gas can be expressed as

$$\frac{pV}{RT_R} = 1 + Ap + Bp^2 + Cp^3 + \cdots = Z \tag{4-6}$$

where A, B, C,··· = virial coefficients and Z = compressibility factor. The compressibility factor

Z illustrates the degree of deviation of the behavior of the real gas, moist air, from the ideal gas

due to the following:

1. Effect of air dissolved in water
2. Variation of the properties of water vapor attributable to the effect of pressure
3. Effect of intermolecular forces on the properties of water vapor itself

For an ideal gas, $Z = 1$. According to the information published by the former National Bureau of Standards of the United States, for dry air at standard atmospheric pressure (29.92 inch. Hg, or 760 mm Hg) and a temperature of 32°F to 100°F (0℃ to 37.8℃) the maximum deviation is about 0.12 percent. For water vapor in moist air under saturated conditions at a temperature of 32°F to 100°F (0℃ to 37.8℃), the maximum deviation is about 0.5 percent.

U. S. Standard Atmosphere

The temperature and barometric pressure of atmospheric air vary considerably with altitude as well as with local geographic and weather conditions. The standard atmosphere gives a standard of reference for estimating properties at various altitudes. At sea level, standard temperature is 15℃; standard barometric pressure is 101.325kPa. Temperature is assumed to decrease linearly with increasing altitude throughout the troposphere (lower atmosphere), and to be constant in the lower reaches of the stratosphere. The lower atmosphere is assumed to consist of dry air that behaves as a perfect gas. Gravity is also assumed constant at the standard value, 9.806 65m/s². Table 4-1 summarizes property data for altitudes to 10,000 m.

Standard Atmospheric Data for Altitudes to 10,000m **Table 4-1**

Altitude, m	Temperature, ℃	Pressure, kPa
−500	18.2	107.478
0	15.0	101.325
500	11.8	95.461
1000	8.5	89.875
1500	5.2	84.556
2000	2.0	79.495
2500	−1.2	74.682
3000	−4.5	70.108
4000	−11.0	61.640
5000	−17.5	54.020
6000	−24.0	47.181
7000	−30.5	41.061
8000	−37.0	35.600
9000	−43.5	30.742
10,000	−50	26.436

Source: Adapted from NASA (1976).

Pressure values in Table 4-1 may be calculated from

$$p = 101.325(1 - 2.25577 \times 10^{-5} Z)^{5.2559} \quad (4\text{-}7)$$

The equation for temperature as a function of altitude is

$$t = 15 - 0.0065Z \quad (4\text{-}8)$$

where Z ——altitude, m

p ——barometric pressure, kPa

t ——temperature, ℃

Equations (4-7) and (4-8) are accurate from −5000 m to 11,000 m. For higher altitudes, comprehensive tables of barometric pressure and other physical properties of the standard atmosphere, in both SI and I-P units, can be found in NASA (1976).

4.2 Psychrometric Chart

Psychrometrics is derived from Greek, in which "psychro" meaning "cold" and "metrics" meaning "measure of." However, HVACR psychrometrics is more than just the measure of cold which is not a true value because all heat is a positive value in relation to no heat, and therefore cold has no number value and is used by most people as a basis of comparison only; it is the science of atmospheric air and water vapor thermodynamics applied to HVACR systems. With an understanding of psychrometrics, one can use psychrometric charts or "psych" charts to illustrate the changes that occur to air as it goes through the air conditioning processes of heating, cooling, humidifying, and dehumidifying. This unit is an introduction to psychrometric charts and describes some of their uses in analyzing HVACR systems. The psychrometric chart is a valuable tool for HVACR engineers and technicians alike.

The psychrometric charts in this unit are for sea level pressures (the standard chart) but psych charts for other elevations are available. Printed or digital psychrometric charts, software programs and Smartphone apps provide the following psychrometric information.

Dry bulb, wet bulb and dew point temperature in degrees Fahrenheit.

Specific volume cubic feet per pound (cf / lb). Density in pounds per cubic foot (lb / cf) is not shown directly on the psychrometric chart but can be calculated as the reciprocal of specific volume.

Enthalpy in Btu per pound of dry air.

Sensible Heat Ratio which is calculated: Btu / hr sensible heat divided by Btu / hr total heat.

Moisture Content (Specific Humidity) which is grains of moisture per pound of dry air or pounds of moisture per pound of dry air.

Relative Humidity as a percentage.

Vapor Pressure in inches of mercury (in. Hg).

Standard Air Dot aka standard air reference point.

Dry Bulb (DB) Temperature

Dry bulb temperature is the temperature of the air measured by a standard thermometer. On the psychrometric chart, the straight vertical lines are dry bulb temperature (example: 60Fdb shown as #1 in Figure 4-1). The dry bulb temperature scale in degrees Fahrenheit is along the bottom of the chart (scale shown from 40Fdb to 110 Fdb). Follow the #1 dry bulb line down to intersect the scale at 60F. Psychrometric charts are available in the following approximate dry bulb temperature ranges: low temperatures −20Fdb to 50Fdb, normal temperatures: 20Fdb to 100Fdb, and high temperatures from 60Fdb to 250Fdb. All charts used in this text are normal temperatures.

Wet Bulb (WB) Temperature

To get wet bulb temperatures above 32℉ use a standard thermometer. Cover the sensing bulb with a wet (distilled water) wick. Expose the thermometer wick to air moving at a velocity greater than 900 feet per minute. The wet bulb temperature scale on thepsychrometric chart is along the 100 percent saturation line on the chart (the curved left line indicated by #3, wet bulb temperatures indicated from 40℉ to 85℉). The wet bulb lines run at about a 30 degree angle across the chart as shown by #2 in Figure 4-1 (40Fwb).

Dew Point (DP) Temperature

Dew point temperature is the temperature at which moisture will start to condense from the air. The scale (shown from 0F to 85F) for dew point temperatures is along the right side of the chart (#7, Figure 4-1). The dew point temperature lines run horizontally, at right angles to the dry bulb temperature lines (#4 in Figure 4-1); in fact, they are the same lines as the specific humidity (also known as humidity ratio) lines (#6, Figure 4-1). For some charts, dew point temperature is along the curved line on the left side of the chart (#8, Figure 4-1). This curved line is the 100 percent saturation line. The dew point is where the humidity ratio line intersects the 100 percent saturation curve.

Specific Volume and Density

Specific volume is the volume of a substance per unit weight. For air, the units of specific volume are cubic feet per pound. Specific volume lines run across the chart at about a 70 degree angle (#10, Figure 4-1). Density and specific volume are reciprocals. Once specific volume is found on the psych chart, density is calculated using the equation: density=1/specific volume. For example, if the specific volume is 13.33 cubic feet per pound then the density is 0.075 pounds per cubic foot (0.075 lb/cf=1/13.33 cf/lb).

Enthalpy

Enthalpy (h) is a thermodynamic property, which serves as a measure of heat energy. Temperature is an indication of heat intensity and Btu is the quantity of heat in a substance. Enthalpy is the measurement of the heat content of a substance used in psychrometrics to find the amount of heat necessary for various processes. On the psych chart, enthalpy represents the energy in one pound of dry air and the grains of moisture associated with it at a given condition. Enthalpy is used in HVACR calculations to find the total heat (typically in Btu per hour) of the air at a given condition. For example, the total heat of the air entering a cooling coil less than the total heat of the air leaving the cooling coil would give the total heat removed by the cooling coil. The enthalpy scale is on the far left side of the psych chart (represented as #9 in Figure 4-1). To find enthalpy on some psych charts extend the wet bulb line from the condition point past the 100 percent saturation line to the enthalpy scale. On other charts (such as represented in Figure 4-1) use a straight edge and draw a line from the left hand enthalpy scale to another enthalpy scale at the bottom or right side of the chart (also shown as #9). The enthalpy line pivots on the condition point. The enthalpy is correct when the line passes through condition point and the enthalpy is the same on the left hand scale and the right or bottom scale. The units of enthalpy are Btu per pound of dry air.

Sensible Heat Ratio

The sensible heat ratio (SHR), also known as sensible heat factor (SHF), is the ratio of sensible heat to total heat (Q_S/Q_t, where Q is quantity). It is also expressed as Btu per hour sensible heat divided by Btu per hour total heat (Btuhs/Btuht). The sensible heat ratio scale is shown on the right side of the chart (#11, Figure 4-1), and is used in conjunction with the SHR reference dot near the center of the chart (#12, Figure 4-1). The SHR reference dot is located at 50% relative humidity and 78Fdb on some charts and 50% relative humidity and 80Fdb on other charts. The sensible heat ratio line is drawn from the SHR dot and extended through the SHR scale.

Moisture Content

Moisture content (w) is the weight of water vapor in each pound of dry air. Moisture content is also known as humidity ratio or specific humidity. The moisture content scale is on the right side of the chart from 0 to 200 grains of moisture per pound of dry air (#5, Figure 4-1). Some psychrometric charts also have a scale that gives the moisture content in pounds of moisture per pound of dry air from 0. 0 to 0. 026 pounds. Specific humidity is the actual amount of moisture in a pound of air. On the psych chart, the specific humidity lines (#6, Figure 4-1) run horizontally at right angles to the dry bulb lines. The units of specific humidity are grains of moisture per pound of dry air or pounds of moisture per pound of dry air. 7000 grains of moisture equals one pound of moisture.

Relative Humidity

Relative humidity (rh) is the ratio of the amount of water vapor, aka moisture, present in the air to the total amount of moisture that the air can hold at a given temperature. Relative humidity is expressed as a percentage. The percent relative humidity lines are the curved lines that start on the left side of the chart at the 100 percent saturation line and decrease going to the right side of the chart. On most psych charts, the designated relative humidity lines are in 10% increments starting at the saturation line. The next line to the right of the saturation line is 90%, then 80%, 70%, etc., with the last line at 10%. In Figure 4-1: #8, indicates relative humidity lines shown at 100% rh, 50% rh and 10% rh. Table 4-2 shows the changes in relative humidity as air temperature changes with a constant amount of moisture content (specific humidity) in the air.

Change in rh with change in air temperature and constant specific humidity. Table 4-2

Specific Humidity	Air Temperature	Relative Humidity
60 gains per pound	100F	21%
60 gains per pound	70F	54%
60 gains per pound	55F	94%

Vapor Pressure

Vapor pressure (Pw) is the pressure exerted by water vapor in the air (#13, Figure 4-1). The units of vapor pressure are inches of mercury (in. Hg).

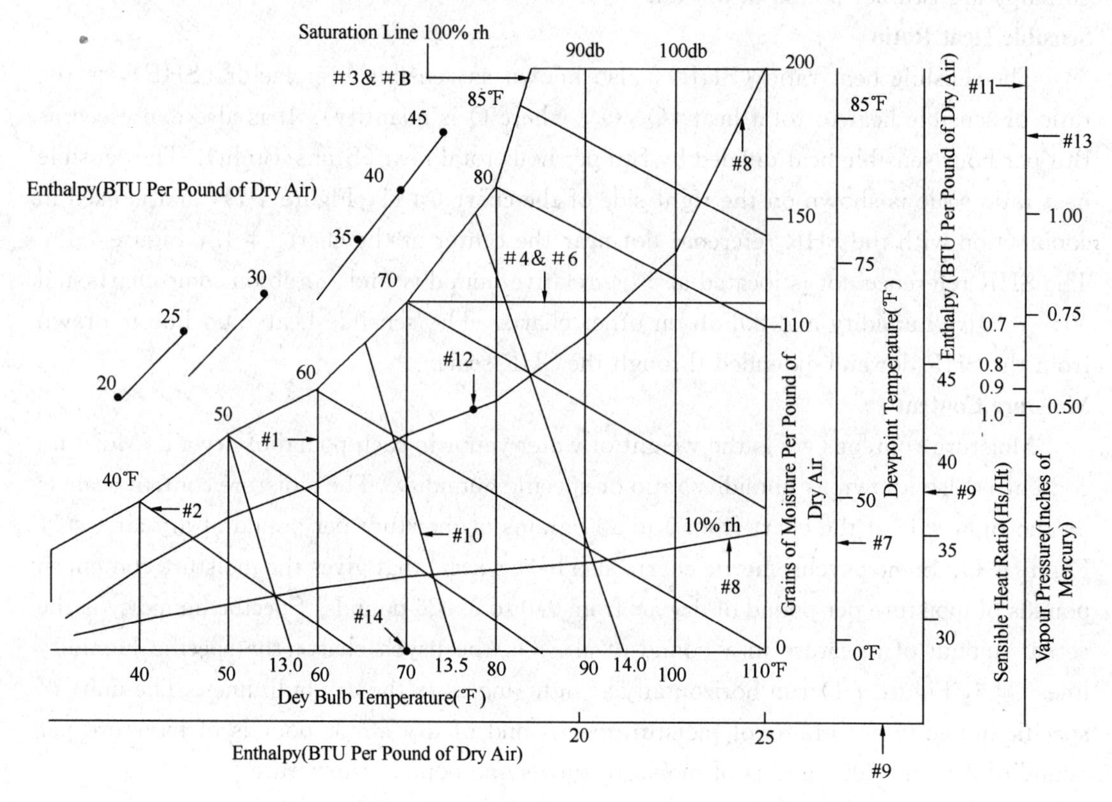

Figure 4-1 Representation of a Psychrometric Chart-Not to Scale

Standard Air Dot

Some psych charts have a dot indicating standard air (#14, Figure 4-1). Standard air is dry air with the following properties: Temperature @ 70F.

Barometric pressure @ 29.92 inches mercury or 14.7 pounds per square inch

Volume (specific volume) @ 13.33 cubic feet per pound

Weight (density) @ 0.075 pounds per cubic foot

4.3 HVACR Process on the Psychrometric Chart

The ASHRAE psychrometric chart can be used to solve numerous process problems with moist air. Its use is best explained through illustrative examples. In each of the following examples, the process takes place at a constant total pressure of 101.325 kPa.

Moist Air Sensible Heating or Cooling

Adding heat alone to or removing heat alone from moist air is represented by a horizontal line on the ASHRAE chart, because the humidity ratio remains unchanged. Figure 4-2 shows a device that adds heat to a stream of moist air. For steady-flow conditions, the required rate of heat addition is

$$q_2 = \dot{m}_{da}(h_2 - h_1) \tag{4-9}$$

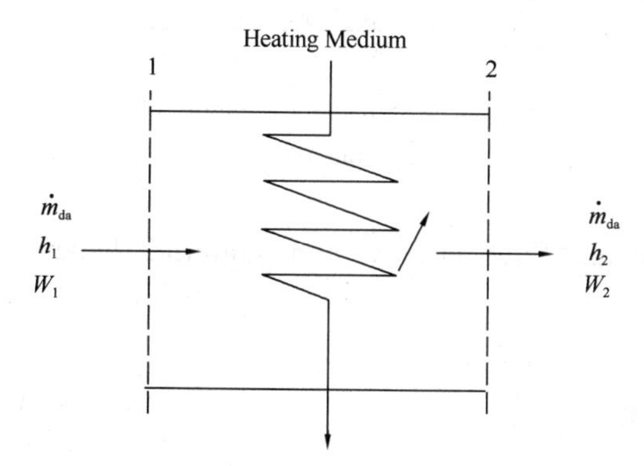

Figure 4-2 Schematic of device for heating moist air.

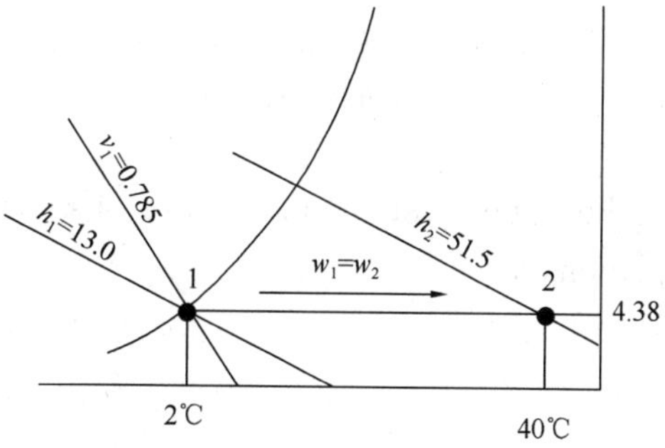

Figure 4-3 Schematic solution for example 4-1.

EXAMPLE 4-1 Moist air, saturated at 2℃, enters a heating coil at a rate of $10\text{m}^3/\text{s}$. Air leaves the coil at 40℃. Find the required rate of heat addition

SOLUTION Figure 4-3 schematically shows the solution. State 1 is located on the saturation curve at 2℃. Thus, $h_1 = 13.0\text{kJ/kg}_{da}$, $W_1 = 4.38\text{g}_w/\text{kg}_{da}$, and $v_1 = 0.785\text{m}^3/\text{kg}_{da}$. State 2 is located at the intersection of $t = 40℃$ and $W_2 = W_1 = 4.38\text{g}_{wT}/\text{kg}_{da}$. Thus, $h_2 = 51.5\ \text{kJ/kg}_{da}$. The mass flow of dry air is

$$\dot{m}_{da} = \frac{10}{0.785} = 12.74\ (\text{kg}_{da}/\text{s})$$

From Equation (4-5),

$$q_2 = \dot{m}_{da}(h_2 - h_1) = 12.74 \times (51.5 - 13.0) = 490\ (\text{kW})$$

Moist Air Cooling and Dehumidification

Moisture condensation occurs when moist air is cooled to a tem-perature below its initial dew point. Figure 4-4 shows a schematic cooling coil where moist air is assumed to be uniformly processed. Although water can be removed at various temperatures ranging from the initial dew point to the final saturation temperature, it is assumed that condensed water is cooled to the final air temperature t_2 before it drains from the system.

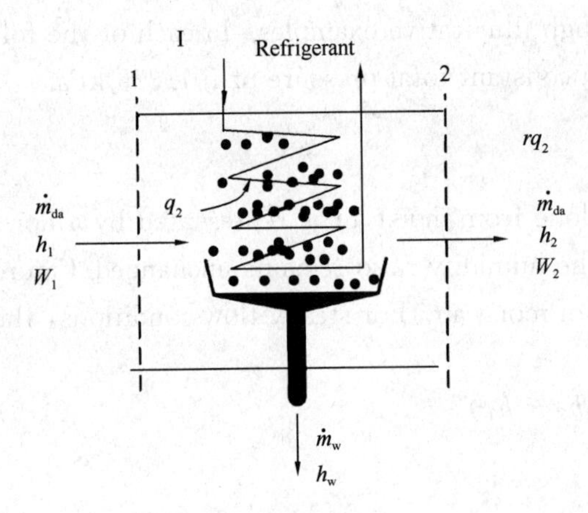

Figure 4-4 Schematic of device for cooling moist air.

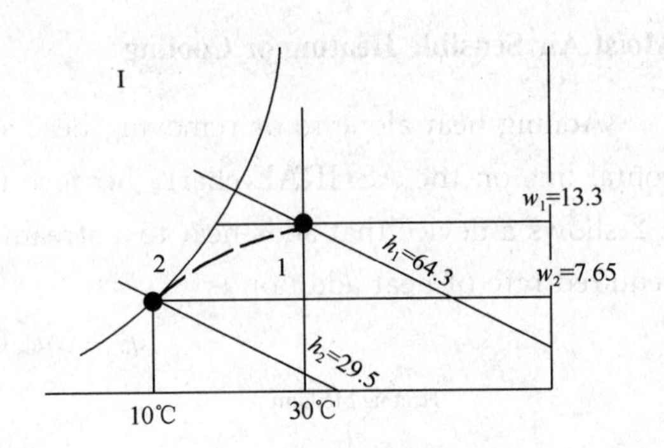

Figure 4-5 Schematic solution for example 4-2.

For the system in Figure 4-4, the steady-flow energy and material balance equations are

$$\dot{m}_{da}h_1 = \dot{m}_{da}h_2 + q_2 + \dot{m}_w h_{w2}$$

$$\dot{m}_{da}W_1 = \dot{m}_{da}W_2 + \dot{m}_w$$

Thus,

$$\dot{m}_w = \dot{m}_{da}(W_1 - W_2) \tag{4-10}$$

$$q_2 = \dot{m}_{da}[(h_1 - h_2) - (W_1 - W_2)h_w^2] \tag{4-11}$$

EXAMPLE 4-2 Moist air at 30℃ dry-bulb temperature and 50% rh enters a cooling coil at 5 m^3/s and is processed to a final saturation condition at 10℃. Find the kW of refrigeration required.

SOLUTION Figure 4-5 shows the schematic solution. State 1 is located at the intersection of t = 30℃ and φ = 50%. Thus, h_1 = 64.3kJ/kg_{da}, W_1 = 13.3g_w/kg_{da}, and v_1 = 0.877m^3/kg_{da}. State 2 is located on the sat-uration curve at 10℃. Thus, h_2= 29.5kJ/kg_{da} and W_2=7.66g_w/kg_{da}. From properties of water at saturation Table, h_{w2}= 42.02kJ/kg_{da}. The mass flow of dry air is

$$\dot{m}_{da} = \frac{5}{0.877} = 5.70\ (kg_{da}/s)$$

From Equation(4-11),

$$\begin{aligned} q_2 &= \dot{m}_{da}[(h_1 - h_2) - (W_1 - W_2)h_w^2] \\ &= 5.70[(64.3 - 29.5) - (0.0133 - 0.00766)42.02] \\ &= 197\ (kW) \end{aligned}$$

Adiabatic Mixing of Two Moist Airstreams

A common process in air-conditioning systems is the adiabatic mixing of two moist airstreams. Figure 4-6 schematically shows the problem. Adiabatic mixing is governed by three equations:

$$\dot{m}_{da1}h_1 + \dot{m}_{da2}h_2 = \dot{m}_{da3}h_3$$

$$\dot{m}_{da1} + \dot{m}_{da2} = \dot{m}_{da3}$$

$$\dot{m}_{da1}W_1 + \dot{m}_{da2}W_2 = \dot{m}_{da3}W_3$$

Eliminating $\dot{m}_{da3}$ gives

$$\frac{h_2 - h_3}{h_3 - h_1} = \frac{W_2 - W_3}{W_3 - W_1} = \frac{\dot{m}_{da1}}{\dot{m}_{da2}} \tag{4-12}$$

according to which, on the ASHRAE chart, the state point of the resulting mixture lies on the straight line connecting the state points of the two streams being mixed, and divides the line into two segments, in the same ratio as the masses of dry air in the two streams.

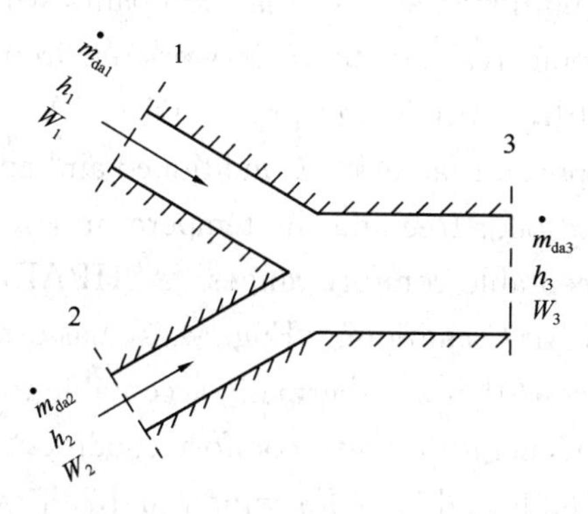

Figure 4-6 Adiabatic mixing of two moist airstreams.

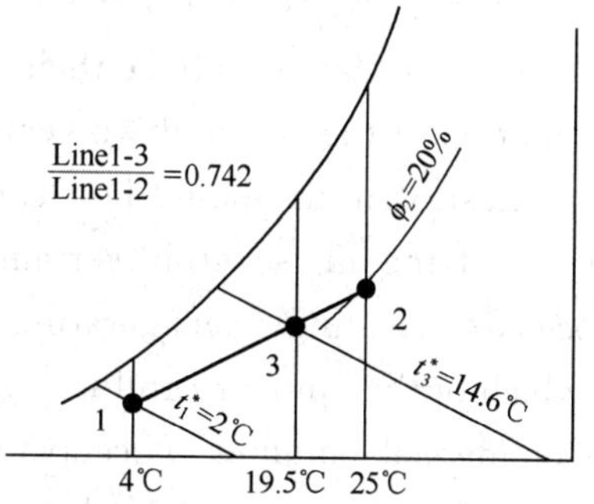

Figure 4-7 Schematic solution for example 4-3.

EXAMPLE 4-3 A stream of 2m³/s of outdoor air at 4℃ dry-bulb temperature and 2℃ thermodynamic wet-bulb temperature is adiabatically mixed with 6. 25m³/s of recirculated air at 25℃ dry-bulb temperature and 50% rh. Find the dry-bulb temperature and thermodynamic wet-bulb temperature of the resulting mixture.

SOLUTION Figure 4-7 shows the schematic solution. States 1 and 2 are located on the ASHRAE chart: $v_1 = 0.789\text{m}^3/\text{kg}_{da}$, and $v_2 = 0.858\text{m}^3/\text{kg}_{da}$. Therefore,

$$\dot{m}_{da1} = \frac{2}{0.789} = 2.535\ (\text{kg}_{da}/\text{s})$$

$$\dot{m}_{da2} = \frac{6.25}{0.858} = 7.284\ (\text{kg}_{da}/\text{s})$$

According to Equation (4-12),

$$\frac{\text{Line 3-2}}{\text{Line 1-3}} = \frac{\dot{m}_{da1}}{\dot{m}_{da2}} \quad \text{or} \quad \frac{\text{Line 1-3}}{\text{Line 1-2}} = \frac{\dot{m}_{da2}}{\dot{m}_{da3}} = \frac{7.284}{9.819} = 0.742$$

Consequently, the length of line segment 1-3 is 0. 742 times the length of entire line 1-2. Using a ruler, state 3 is located, and the values $t_3 = 19.5$℃ and $t_3^* = 14.6$℃ found.

Unit 5 Introduction to HVACR Systems

Properly designed, installed, tested and maintained HVACR systems provide conditioned air for the work process function, occupancy comfort, and good indoor air quality while keeping system costs and energy requirements to a minimum. Commercial HVACR systems provide building work areas with "conditioned air" so that occupants will have a comfortable and safe work environment. People respond to their work environment in many ways and many factors affect their health, attitude and productivity. "Air quality" and the "condition of the air" are two very important factors. "Conditioned air" and "good air quality," mean that air should be clean and odor-free and the temperature, humidity, and movement of the air is within certain acceptable comfort ranges. ASHRAE, the American Society of Heating, Refrigerating and Air Conditioning Engineers, has established standards which outline indoor comfort conditions that are thermally acceptable to 80% or more of a commercial building's occupants. Generally, these comfort conditions, sometimes called the "comfort zone," are between 68°F and 75°F for winter and 73°F and 78°F during the summer. Both these ranges are for room air at approximately 50% relative humidity and moving at a slow speed (velocity) of 30 feet per minute or less.

The HVACR system is simply a group of components working together to provide heat to, or remove heat from, a conditioned space. For example the components in a typical roof-mounted package unit HVACR system, shown in Figure 5-1, are:

Figure 5-1 Roof-mounted Unit (aka rooftop unit, RTU). Natural gas heating and vapor compression DX cooling.

1. An indoor fan (aka blower) to circulate the supply and return air.
2. Supply air ductwork in which the air flows from the fan to the conditioned space.
3. Air devices such as supply air outlets and return air inlets.

4. Return air ductwork in which the air flows back from the conditioned space to the unit.

5. A mixed air chamber (aka mixed air plenum) to receive the return air and mix it with outside air.

6. An outside air device such as a louver, screened opening or duct to allow for the entrance of outside air into the mixed air plenum.

7. A filter section to remove dirt, debris and dust particles from the air.

8. Heat exchangers such as a refrigerant evaporator coil and condenser coil for cooling, and a furnace for heating.

9. A compressor to compress the refrigerant vapor and pump the refrigerant around the refrigeration system.

10. An outdoor fan (aka blower) to circulate outside air across the air-cooled condenser coil.

11. Controls to start, stop and regulate the flow of air, refrigerant, and electricity.

The abbreviation HVACR is certainly a mouthful, and so it is not unusual to ask the question, "What does this mean, and how does it impact me?" However, the answer is not so simple, and a standard definition may not explain very much. This is because the HVACR industry is a complex network that our entire society relies on more today than ever before. Just think how your world would change without refrigeration for your food or drinks and without air conditioning in your car or classroom. Try to visualize how this would affect the greater population, from food distribution networks, to hospital care, to housing for the elderly. As a trained and skilled HVACR technician, you can make a positive impact on society. You can contribute to this growing industry to ensure that systems work efficiently and safely and are environmentally friendly.

5.1 History and Overview of HVACR

Heating

In an attempt to better understand HVACR, let's break it down component by component. The **H** for heating seems easy. The history of heating a space by burning wood starts in our earliest times and continues to the present. Elaborate systems using firewood heated Roman buildings. Channels were built underneath the floors to draw heat from a fire, thus warming the building and creating the first central heating systems.

Wood, peat, and coal remained the primary heating fuels for centuries. Many early buildings had open fireplaces. But fireplaces are an inefficient way of heating because too much of the heat produced is drawn up the chimney. Although early seventeenth-century European masonry-type stoves burned wood safely at high efficiency, the next major step in heating technology in America was the metal stove. Benjamin Franklin is credited with inventing a castiron stove that was several times more efficient than any other stove at that time. Many people still use decorative, efficient stoves to provide much, if not all, of their heating needs as shown in Figure 5-2.

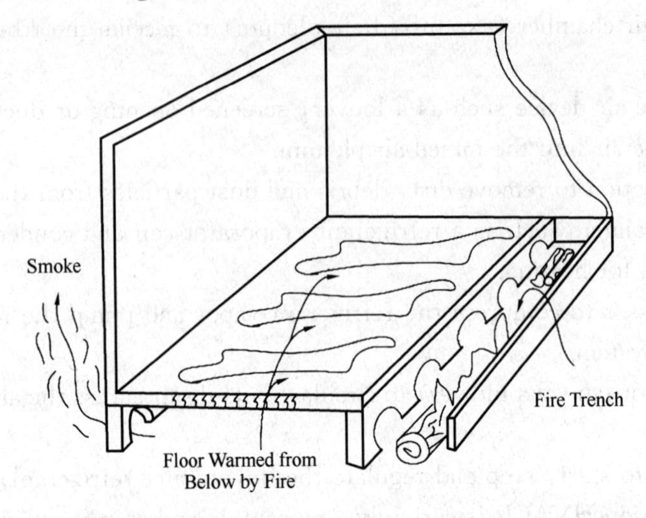

Figure 5-2 Romans used fires channeled below floors as early heating systems.

However, wood heat is only one alternative, because today there are many more choices for heating. Gas heat, oil heat, electric heat, and solar heating systems are common. Heat pumps that use a refrigeration system for heating can be very efficient. Geothermal heating systems that utilize the heat from within the earth are becoming more popular. New, environmentally friendly ideas and efficient designs are continually being developed, tested, operated, and maintained by people just like you entering the industry. So you can see that just the **H** alone is a large and important sector.

Ventilation

Next comes the **V** for *ventilation.* Before the invention of chimneys, fires were burned in the center of a room with smoke having to escape through holes in the roof. When early homes were heated by wood fires, the smoke would permeate the entire building. Although people were warm, the health hazards from this smoke exposure were harmful. As an improvement, early Norman fireplaces in England were designed to allow the smoke to escape through two holes in the side of the building. It was obvious that something needed to be done to improve the air quality.

A properly ventilated building allows for the air to flow and exchange so that harmful particulates such as those in smoke are not allowed to accumulate. Fresh air also brings oxygen into the space, but it becomes depleted over time. A simple ventilation system can consist of only a fan and some minor ductwork for transporting the air. More complex systems circulate air throughout entire buildings through a vast network of ducts and blowers.

Air Conditioning

The AC stands for *air conditioning*. Generally this is considered by most people to be a way to cool a space, but as you will learn, this term encompasses much more. Artificially cooling the air in a living space dates back to the earliest centuries. In ancient Greece, large wet woven tapestries were hung in natural drafts so that the air flowing through and around the tapestries was cooled by the evaporating water. As the water evaporated it would remove heat, just like when you perspire to remain cool shown in Figure 5-3. Some manufacturers sprayed water in factories for cooling as early as the 1720s. Evaporative cooling is still used extensively in residences and businesses throughout the southwestern United States, where typical summer conditions are very hot and dry.

Ice was the primary means of cooling air for many years. The Romans packed ice and snow between double walls in the emperor's palaces. John Gorrie patented the first mechanical air-conditioning system in 1844. His system was used to cool sick rooms in hospitals in Florida. The United States Capitol building in Washington, DC, was first air conditioned using ice in 1909. Rumor has it that when the legislators got really involved in controversial debates, more ice was required to keep the building cool. The phrase "tons of air conditioning" we use today came from this era in history, when tons of ice were used for cooling as illustrated in Figure 5-4.

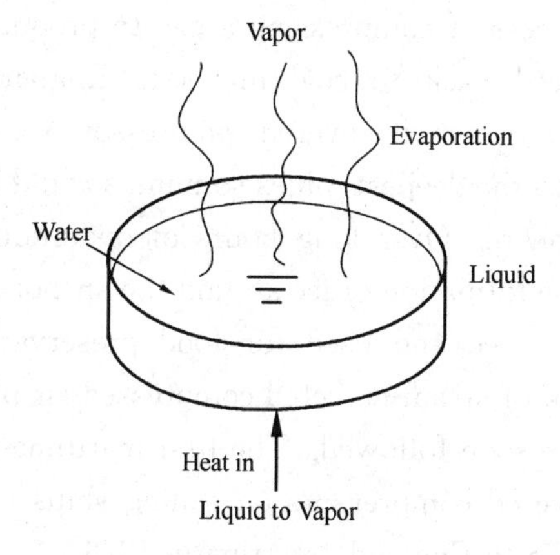

Figure 5-3 When water evaporates, heat is absorbed. This change of state is also referred to as a phase change.

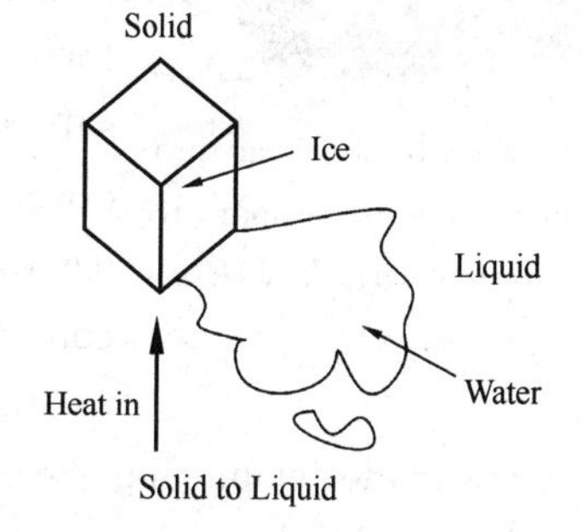

Figure 5-4 When ice melts, heat is absorbed.

Refrigeration

Finally, the R stands for refrigeration, which is a necessary component for most air-conditioning systems; however, refrigeration systems are more commonly considered to be used for keeping food cold. That is why very often you may see the abbreviation HVACR, which implies air conditioning only. The broader term HVACR includes both air conditioning and refrigeration systems.

The first use of refrigeration was for the preservation of food. Ice was harvested from frozen lakes and stored for later use. Sometimes it could be kept all summer long in ice houses. Ice harvesting remained a flourishing industry well into the twentieth century.

Archeologists have discovered that the first evidence of man making ice appeared more than 3,000 years ago, about 1,000 BC. Peoples living in northern Egypt, the Middle East, Pakistan, and India made ice using evaporation. Archeological excavations in these regions have discovered ice-producing fields that covered several acres. The ice was produced in shallow clay plates, about the size of a saucer. The water in these clay plates wept through the clay. This water dampened the small straw mats holding the clay plates in racks a few feet above the ground. The straw aided evaporative cooling of the water. Under the right conditions of temperature and humidity, a thin film of ice would form overnight on each clay plate.

Figure 5-5 Snowblowers can produce artificial snow by evaporative cooling. (*Courtesy of Red River Ski Area*)

Producing ice in this way is also the principle behind modern snow-making equipment. A snow-producing machine like the one in Figure 5-5 can make snow by evaporative cooling even when the temperatures on the ski slopes are above freezing.

Today, a majority of refrigeration systems use what is referred to as mechanical vapor compression. The mechanical process of compressing a gas to produce cooling can be traced back to coal mines in England. Large steam-driven or water-powered compressors were used to force air into the deepest mines so miners could work in a safe atmosphere. Over long hours of operation, miners observed the formation of ice around the air nozzles. This ice was collected and used for food preservation. The construction of steam-powered compressed-air plants that produced ice soon followed. The first maritime refrigeration units were made by putting steam-powered compressors on sailing ships to make it possible for beef to be shipped from Australia to England, starting in 1876.

Refrigerant capacity is measured in tons. One ton of capacity is equivalent to the amount of heat that 2,000 lb of ice can absorb in one day. The amount of latent heat required to change 1 ton of ice into 1 ton of water is 288,000 BTU. If this amount is divided by 24 hr per day, the equivalent is 12,000 BTU/hr.

HVACR and the Refrigeration Cycle

Now that you have a better understanding of what HVACR means, it is easy to see that it encompasses a broad spectrum of needs and applications. Although the methods for heating can vary considerably, the majority of cooling applications are based on the refrigeration cycle. When ice changes to water, heat is absorbed, which makes ice a viable refrigerant. But ice is hard to store and takes up a lot of space. Water is easier to use because it can be pumped and doesn't need the insulation that ice requires. When water evaporates to vapor it also absorbs heat, but then the water needs to be replaced, and this uses up a lot of water over time.

If the vapor can be recovered and turned back into water, then this cycle reduces the total amount of water needed. Even so, the major disadvantage with this type of evaporative cooling is that the lowest temperature that can be reached is dependent on the properties of water.

Notice that with both ice and water, it is their change of state that allows for heat to be absorbed. It is this important principle that serves as the basis for most refrigeration systems today, but instead of using water, other fluids with different properties and lower boiling points, called refrigerants, are now used. This allows for much colder temperatures, far below freezing. The "refrigeration cycle" therefore continually evaporates and condenses refrigerants to absorb and then throw away the heat.

A compressor is used like a pump to raise the pressure and circulate the refrigerant through the system shown in Figure 5-6. A condenser is used to remove heat from the refrigerant as it turns into a liquid. An expansion device drops the pressure to allow the refrigerant to change back from liquid to vapor in the evaporator. Heat is absorbed in the evaporator and then thrown away in the condenser. The refrigerant does not wear out and circulates around and around during operation. Most refrigeration systems in use today operate using this type of cycle.

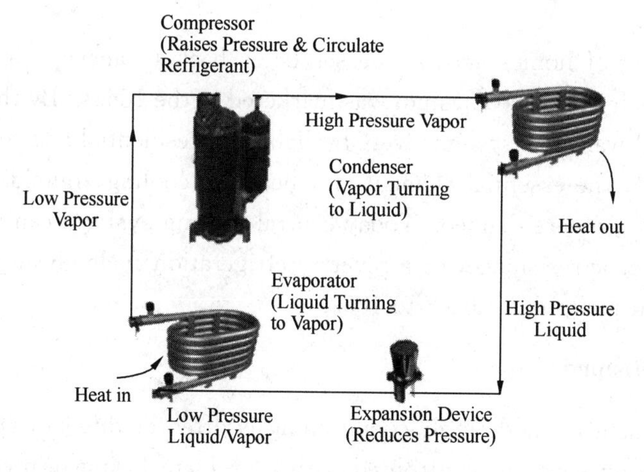

Figure 5-6 The basic refrigeration cycle consists of four major components: compressor, condenser, expansion device, and evaporator.

5.2 Today's Heating, Air conditioning, and Refrigeration

"Environmental heating and air conditioning" refers to the control of a space's air temperature, humidity, circulation, cleanliness, and freshness, and it is used to promote the comfort, health, and/or productivity of the inhabitants. Homes, offices, schools, colleges, factories, sporting arenas, hotels, cars, trucks, and other vehicles such as aircraft and spacecraft are heated and cooled. The main purpose of environmental heating or cooling is to help maintain the body temperature within its normal range. Generally, the term air conditioning is used when the space temperature is above 60°F (15℃), and refrigeration is the term used when the space temperature is below 60°F (15℃).

Process heating and cooling are used to aid in manufacturing or to keep equipment at a desired temperature. An area used to process meat or vegetables may be cooled to help preserve the product. Computer rooms are cooled so the equipment lasts longer and is able to stay online due to the heat being removed from the space. Computers would not operate properly if heat was not absorbed from the space. Remote pumping stations may be heated to prevent pipes from freezing. The main purpose of process heating or cooling is to maintain the temperature of things or processes within their required range.

Without our ability to control the environment, it would be impossible for us to explore space or the bottom of the ocean or even to enjoy the comfort of a transcontinental jet ride at 35,000 ft. So our ability to control our environment has served both to improve the quality of life and to enhance our scientific endeavors.

An operating room is cooled to aid with the surgery as well as for the comfort of the patient or surgeon. Therefore, an operating room is an example of process cooling even though it may be within the normal air-conditioning temperature range.

Modern Heating

Central heating of homes and businesses dates back to ancient times, but the first commercial warm-air fan-driven system was marketed in the 1860s. By the 1900s a number of different central warm-air systems were available for residential and commercial applications, and in 1908, the essential elements for heating, cooling, humidifying, dehumidifying, and filtering air were defined. Today central heating systems can use warm air, hot water, steam, electric resistance, or a reverse refrigeration cycle (heat pump). The basic theory for the heat pump dates back to 1852.

Modern Air Conditioning

The development of modern air conditioning is often credited to Dr. Willis Carrier. Dr. Carrier, an engineer, was confronted with a problem facing printers. As paper was printed with one color, the dampness in the ink caused the paper to stretch slightly, and it was nearly impossible for the second color to be printed without being misaligned. Dr. Carrier determined that a means for controlling the humidity was necessary and developed the first air-conditioning system for the printing industry. His invention, called an "Apparatus for Treating Air," was patented in 1906 shown in Figure 5-7. His invention quickly found favor not only for dehumidifying but also for cooling. Through the 1940s and 1950s, businesses would proudly display signs reading "Air Conditioned."

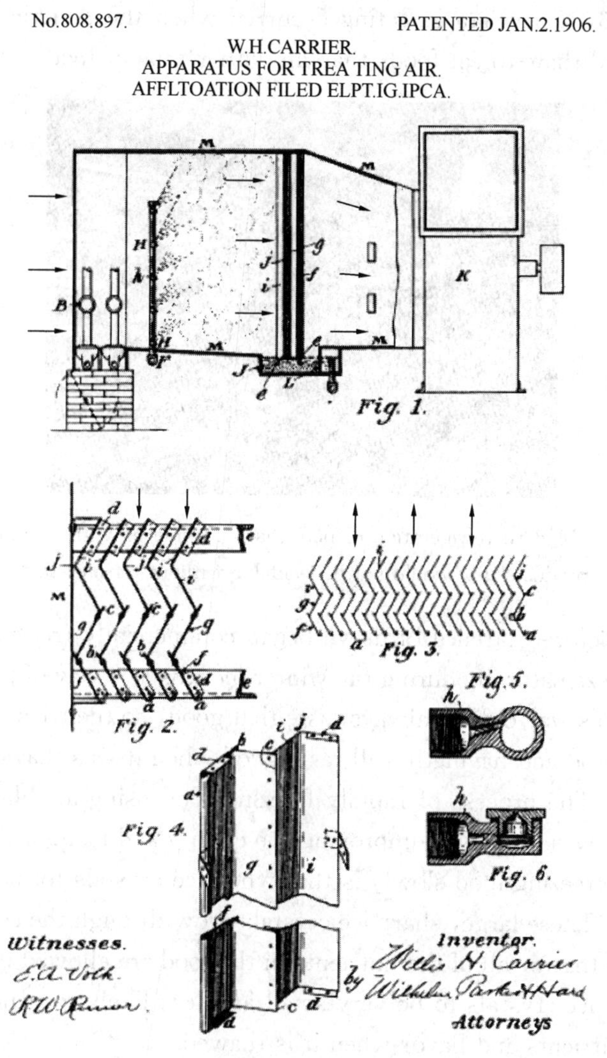

Figure 5-7 The patent for the first apparatus for cooling air, invented by Willis Carrier.

Dr. Carrier designed the psychrometric chart as we know it today. (This chart displays the properties of air, such as temperature, humidity, and volume, and is commonly used for many HVACR applications.)

Mass air conditioning of homes began in the late 1950s with window air conditioners. Central residential air conditioning started to become popular in the mid-1960s. Today most of us cannot imagine living in a home anywhere in the country that does not have air conditioning.

Modern Refrigeration

Clarence Birdseye made another major contribution to the industry. He developed the process of freezing foods in 1922. Today, supermarket freezer displays provide us with a variety of food products that would not be possible to preserve any other way as shown in

Figure 5-8. In 2006, a new era in eating occurred when the American public purchased more heat-to-eat and thaw-to-eat foods than any other type of food.

Figure 5-8 Modern refrigeration display cases provide us with a variety of food products that would not be available without refrigeration.

- **Frozen foods** Before Clarence Birdseye began commercially freezing food, people had allowed food to freeze naturally during the winter months as a way of preserving it for later use. Food frozen this way did not always taste that good, so the trick was to come up with a way of freezing food and having it still taste good when it was thawed.
- **Quick freezing** The process of rapidly freezing food using air blast, contact, and/or immersion freezing was the key to improving the quality and taste of thawed frozen foods. The problem with freezing food slowly is that when ice crystals form over time, they become much larger. These large, sharp ice crystals grow through the cell walls of the food, and when the food thaws, all of the nutrients in the food are allowed to drain away. Quick freezing causes the ice crystals to be very small and less likely to penetrate cell walls, so the food retains nutrients and flavor when it is thawed.

5.3 Employment Opportunities

The HVACR industry represents one of the largest employment occupations in the country. Our industry, for example, is one of the largest consumers of electric and gas utilities in the nation. More electricity and natural gas is consumed producing heating and cooling than for any other single use. The size of the industry has been growing steadily since the late 1960s, when residential central systems became popular. The installation and servicing of HVACR systems will always be an expanding occupation. No one builds a home or business without some type of heating and/or cooling system, which requires designing, installing, and servicing by skilled and trained technicians.

Residential Air Conditioning and Heating

Most residential heating systems have a heating capacityof 50,000 to 150,000 BTU/

hr. The majority of residential air conditioning systems are 5 tons or less. Both of these sizes will obviously vary greatly, depending on the region of the country you are working in. In addition, there are many very large homes being built, requiring systems that could easily be classified as light commercial because of their size and/or complexity.

To protect the public from potentially dangerous individuals, some businesses and/or local and state governments require criminal background checks on anyone involved with in-home service work. These checks may go back 25 years or more into an individual's past. Check with your local or state governmental department that regulates in home service work if you feel there is something in your past that might affect your ability to work in residential service. In most states these checks are only required for in-home service work, so you may still be able to work in new construction or in the commercial or industrial areas.

Commercial Air Conditioning and Heating

The term *commercial* is used to refer to any system that is used in commercial buildings (for business) that provides cooling or heating. These systems may be as small as a fraction of a ton in size to several thousand tons in cooling capacity and/or from 1,000 BTU/hr to hundreds of thousands of BTU/hr. Commercial systems may be operated independently of any other system or be integrated with a building automation system. Because of the vast differences in the types of equipment and system complexity, commercial technicians often specialize in a single type of system or group of systems.

Commercial and Industrial Refrigeration

The terms *commercial refrigeration and industrial refrigeration* are applied to retail food and cold-storage equipment and facilities. Examples of commercial equipment and systems include refrigeration equipment found in supermarkets, convenience stores, restaurants, and other food service establishments. Industrial refrigeration can include long-term storage either as cold storage or medium or low temperature refrigeration systems that are generally larger scale operations.

Types of Jobs

There are a variety of occupational specialties offered within the HVACR industry. These occupations range from the basic entry-level helper to the systems designer. Although the work involved with heating, air conditioning, or refrigeration equipment and systems is similar in theory, there is a significant difference between the work done in the areas of residential, light commercial, commercial, and industrial. These areas of heating, air conditioning, and refrigeration generally relate to the size (capacity) and complexity of the system. However, technicians may find the exact same equipment used in one home being used in a commercial shop or factory. In these cases the distinguishing factor is whether you are working in someone's home or in a business.

• **Entry-level helper** The entry-level helper (firstyear apprentice) provides the senior technician with assistance installing and servicing equipment. Most medium and large mechanical contracting companies use a number of helpers to assist with the installation and service of residential and commercial systems. A helper may be expected to assist in lifting, carrying, or placing equipment or components. He or she may also run errands to pick up parts and clean up the area following installation or service. Helpers receive basic safety training, and if they will be driving, they must have good driving records.

• **Rough-in installer** The initial installation process is referred to as rough-in. In this process the technician (first-through third-year apprentice) will install the refrigerant lines, electrical lines, thermostat and control lines, duct boots, and duct run and set the indoor and outdoor units. The rough-in technician must have an understanding of duct layout, blueprint reading, and basic hand tools and good brazing skills.

• **Start-up technician** Once the system has been installed and all of the components are ready for operation, a start-up technician (fourth-year or fifth-year apprentice) will go through the manufacturer's recommended procedures to initially start a system. Because much of the HVACR system has been field installed, this checkout procedure is essential to ensure safe and efficient operation. The start-up technician records all of the information requested by the manufacturer's warranty. Start-up technicians must be skilled with electrical troubleshooting and refrigerant charging and have good reading comprehension and writing skills.

• **Service technician** The service technician (fourth-year apprentice to journeyman) is the individual who provides the system owner with repair and maintenance. Service technicians are the people who must be able to diagnose system problems and make the necessary repairs. Service technicians must be skilled in diagnosing electrical problems, refrigerant problems, and air-distribution problems. Technology has enabled the field tech to stay in close contact with his service manager. This allows the highly experienced service managers to provide assistance to technicians as they come upon new problems. The technician can also call upon the office to research unique problems to determine the best, most efficient way of making the repair.

• **Sales** HVACR sales are divided into two major categories: inside sales and outside sales. Inside sales deal primarily with system sales to other air conditioning contractors. Outside sales may be to both contractors and end users. Working in outside sales or consumer sales requires the technician to have a good understanding of cost and value of equipment so that the owner can make an informed choice.

• **Equipment operator** Equipment operators are required by local ordinance and state law to be present anytime large central heating and air-conditioning plants are in operation. Their primary responsibility is to ensure the safe and efficient operation of these large systems. They must have a good working knowledge of the system's mechanical, electrical, and computer control systems to carry out their job. They sometimes need to hold a city or state license to become an operator. Equipment operators generally work by themselves or as part of a small crew. They often are required to have good computer skills when buildings have computerized building-management systems.

- **Facilities maintenance personnel** Facilities maintenance personnel are responsible for planned maintenance and routine service on systems. They may work at a single location or have responsibilities for multiple locations, such as school systems. Facilities-maintenance personnel typically maintain systems and provide planned maintenance. They may work alone or as part of a crew, depending on the size of the facility. Maintenance personnel may from time to time have duties and responsibilities outside of the HVACR trades, such as doing minor electrical plumbing and carpentry projects for the upkeep of the building.
- **Service manager** A service manager is typically a skilled HVACR technician with several years of experience. This individual oversees the operation of a company or maintenance department. He or she must have good management skills, communication skills, and technical expertise. Service managers typically assign jobs to other technicians and employees. They must then oversee these individuals' jobs.
- **Systems designer** For small buildings, contractors normally size and select HVACR systems and equipment. There are many industry-standard sizing and design guides available from trade associations such as the ACCA (Air Conditioning Contractors of America). For larger buildings, mechanical, architectural, or building services engineers may be required by law to design and specify the HVACR systems. Specialty mechanical contractors will work with the design plans to build and commission these systems.

Unit 6 Refrigerating Systems

The purpose of any refrigeration system is to move heat. Typically the heat is being moved from a place where it is not wanted to a place where it is unobjectionable. Most often this involves moving heat from a cool place to a warmer place. Pushing heat uphill like this requires energy. There are many types of cooling systems, including absorption, evaporative, thermoelectric, and mechanical compression. The different types of refrigeration systems differ in how they accomplish this task, but they all involve using energy to move heat.

6.1 Mechanical Compression Refrigeration

The mechanical compression-refrigeration cycle is by far the most common. The mechanical compression refrigeration cycle is applied in air conditioning, commercial refrigeration, and industrial process refrigeration. Air conditioning systems cool people. They use refrigeration to provide comfort cooling and dehumidification in residential and commercial buildings. Commercial refrigeration systems are chiefly concerned with cooling products. Commercial refrigeration systems can be found in grocery stores, restaurants, refrigerated warehouses, and science labs. Industrial process refrigeration is used to cool equipment and machinery in large industrial manufacturing plants.

The heat being moved can often be used. For example, heat rejected by an air-conditioning system can be used to heat hot water. Using the heat that is removed from the area being cooled to heat domestic hot water reduces the energy required to heat the water, improving the overall system efficiency.

Compression-Cycle Operating Principles

The transfer of heat in the compression-refrigeration cycle is performed by a refrigerant operating in a closed system. Most machines that cool or refrigerate something use the same mechanical refrigeration cycle. This cycle depends upon a few basic physical principles. Some of these principles are intuitive, such as that heat travels from hot to cold. Other principles are less intuitive, such as understanding that boiling is a cooling process. Understanding the following handful of basic physical principles involved makes understanding the mechanical refrigeration cycle much easier.

- the transfer of heat in the compression-refrigeration clack of heat.
- Heat travels from hot to cold.
- Liquids absorb large amounts of heat when they boil off to a gas.

- Gases give off large amounts of heat when they condense to a liquid.
- When something is boiling or condensing, its temperature remains the same.
- The temperature at which a liquid boils is controlled by its pressure.

Refrigeration machines do not pump cold in; they pump heat out. Imagine that your air conditioner is like a pump on the bottom of a ship. The purpose of the pump is to pump water out, preferably faster than it leaks in. An air conditioner has a similar purpose, to pump heat out faster than it leaks into your house. The problem is that heat only wants to travel from hot to cold, and when you need your air conditioner it is usually hotter outside than inside. So the challenge is to pump heat from inside, where it is relatively cool, to outside, where it is relatively hot.

Remember that boiling is a cooling process! When a liquid boils, it absorbs large amounts of heat in much the same way that a sponge absorbs water. When you place a dry sponge on a puddle, the sponge turns soggy. The puddle ends up drier, but the sponge ends up wetter. A boiling liquid acts like a sponge soaking up heat. As the liquid boils it takes heat away from whatever it is touching, leaving the surrounding area cooler. When water is poured on a hot frying pan, clouds of steam billow forth. The energy comes from the frying pan to produce the steam, cooling off the frying pan. The reason people think of boiling as hot is because water boils at 212°F under normal atmospheric pressure, and that is hot. But suppose water boiled at 50°F instead? People might pour boiling water over themselves in the summer to help cool off. It actually is possible to boil water at 50°F! Remember that a liquid's boiling temperature is controlled by its pressure. Water will boil at 50°F if placed under a low enough vacuum.

Refrigerant, like water, can boil at almost any temperature by controlling its pressure. The refrigerant in an air conditioner boils in the cooling coil at a low pressure. The low pressure keeps the refrigerant boiling temperature below the temperature of the air being cooled. The refrigerant temperature stays low even as it absorbs heat because the temperature of a boiling liquid remains stable. This is how heat is absorbed from the space being cooled, by boiling the refrigerant at a low pressure and temperature. During this process, the refrigerant changes from a liquid to a gas. This is done in a component called an evaporator because the refrigerant is evaporating. Heat flows from the warm air traveling over the evaporator to the cold refrigerant boiling inside the evaporator.

The compressor pumps gas from the low-pressure evaporator to the high-pressure condenser. The word compress means to squeeze. It increases the gas pressure by squeezing the gas into a smaller space. It is important to note that the compressor is designed to pump only gas. Liquid does not squeeze very well. Putting liquid through the compressor will damage it.

When the gas is squeezed into a smaller space, both the pressure and temperature rise. The high pressure raises the refrigerant temperature above the hot outdoor air temperature. Now the refrigerant is sent outside to be condensed at a high temperature and pressure. The heat that was absorbed in the evaporator is expelled outside by condensing the refrigerant back to a liquid. The temperature of the refrigerant stays high even as the refrigerant loses heat because temperature remains constant during condensation. Heat travels from the hot refrigerant to the outdoor air because the air is cooler than the refrigerant. During this process, the refrigerant changes from a gas to a liquid. This is done in a component called a condenser because the refrigerant is condensing. Notice that in both the evaporator and condenser, heat is going from hot to cold. The trick is keeping the refrigerant pressures low in the evaporator where heat is absorbed and high in the condenser where heat is expelled.

The high pressure and temperature of the refrigerant must be reduced before sending it into the evaporator to boil at a low pressure and temperature. The metering device does this. The metering device is basically a small restriction that limits the amount of refrigerant entering the evaporator. Limiting the flow of refrigerant increases the pressure before the metering device and decreases the pressure after the metering device. The drop in pressure causes the refrigerant to also drop in temperature. This cold refrigerant is now ready to absorb heat in the evaporator.

System High Side and Low Side

For all practical purposes there are two pressures in the system: the low-side pressure and the high-side pressure. The compressor and the metering device work in partnership to maintain this pressure difference. Figure 6-1 shows how the compressor and metering device, the expansion valve, divide the system into the high- and low-pressure sides.

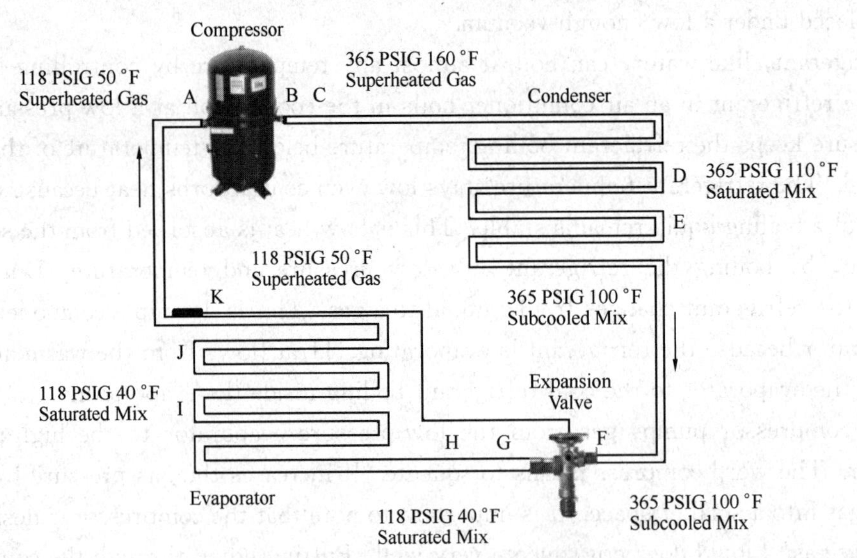

Figure 6-1 The compressor and metering device divide the system into a high-pressure side and a low-pressure side.

The low side contains the low-pressure liquid and vapor refrigerant and is the side that absorbs heat. The metering device controls the flow into the evaporator; the expansion of the refrigerant causes a pressure drop. The low side of the system starts at the metering device and extends through the evaporator and the suction line up to the compressor inlet.

The high side contains the high-pressure vapor and liquid refrigerant and is the part of the system that rejects heat. The compressor pumps the refrigerant into the condenser and maintains the high pressure. The high side of the system starts at the compressor outlet and continues through the discharge line, condenser, liquid line, and up to the metering device.

The body is cooled by evaporation of the perspiration on the skin. The evaporation is reduced when the air is at a higher relative humidity, so the body's cooling mechanism does not work as well. That is why you feel more comfortable at 80°F with 40 percent relative humidity as compared to 75°F and 90 percent relative humidity. To be effective, air-conditioning systems must remove both heat and humidity.

Refrigeration System Relationships

There are key relationships within the refrigeration cycle. Understanding these will improve your understanding of the refrigeration system. The most obvious is the link between pressure and temperature. If the refrigerant is a high pressure, it is also a high temperature. If it is a low pressure, it is also a low temperature. Another important link is between the refrigerant state and its heat content. Gas always hasa relatively high heat content, and liquid always has a relatively low heat content. The heat content of the refrigerant is affected much more by the state of the refrigerant than by its temperature. This is why the refrigerant can have a low temperature and a high heat content leaving the evaporator. Refrigerant will be found in the system in one of three conditions: superheated, saturated, or subcooled. It will be superheated anywhere it is all gas, subcooled anywhere it is all liquid, and saturated anywhere it is changing from one to the other. The refrigerant is superheated leaving the evaporator, through the compressor, and up to the condenser. It is sub-cooled leaving the condenser up to the expansion device. It is saturated in both the condenser, where it is changing from gas to liquid, and in the evaporator, where it is changing from liquid back to gas.

Each of the four main components has an opposite. The compressor and expansion device are opposites. The compressor changes a low-pressure and temperature gas into a high-pressure and temperature gas by reducing its volume. The expansion device changes a high-pressure and temperature liquid into a low-pressure and temperature liquid by expanding its volume. Together they maintain the pressure difference in the system. The condenser and evaporator are opposites. The condenser changes a high-pressure and temperature gas into a high-pressure and temperature liquid, losing a large amount of heat in the process. The evaporator turns a low-pressure and temperature liquid into a low-pressure and temperature gas, gaining a lot of heat in the process.

Heat Flow in the Refrigeration Cycle

Heat is added to the refrigerant in the evaporator. Heat from the product load is transferred to the refrigerant as it boils in the evaporator. This is the majority of heat absorbed by the refrigerant. A small heat gain occurs in the piping from the evaporator up to where the refrigerant enters the compressor. Heat gain in the suction line is not desirable and can be reduced by insulating the suction line. The compressor adds a sizable quantity of heat to the refrigerant. This heat is equivalent to the work done in compressing the refrigerant. In a suction gas cooled semi-hermetic or hermetic motor compressor unit, the motor heat is also transferred to the refrigerant.

The heat added in the evaporator and by the compressor is removed in the condenser. The heat is transferred from the refrigerant in the condenser to the air or water flowing over the condenser. The heat balance of the overall system is shown in the following formula:

Heat added in the evaporator + heat of compression = heat rejected in the condenser.

6.2 Other Types of Refrigeration Systems

Absorption Refrigeration Systems

Absorption refrigeration systems use heat and the process of absorption instead of a compressor to move the refrigerant. The absorption process takes advantage of the fact that some chemicals have an affinity for other chemicals. One of the most common combinations is to use ammonia as the refrigerant and water as the absorbent. This combination is used in smaller absorption systems. Large absorption chillers use water as the refrigerant and lithium bromide as the absorbent. Absorption systems really have two cycles: a refrigerant cycle and an absorbent cycle. Some systems use a third gas, typically hydrogen, to help the refrigerant move from the low side of the system back to the high side without using any mechanical devices. Other systems use a solution pump to move solution from the low side to the high side. This discussion will focus on ammonia-water absorption systems used for residential chillers. Figure 6-2 shows a simplified drawing of an ammonia-water absorption cycle.

Ammonia is a poisonous and somewhat flammable gas! Do not attempt to service an ammonia system unless you have received specific training in the safe handling of ammonia refrigerant.

The transfer of heat is essentially the same in the absorption system as in the traditional mechanical compression cycle. Heat is absorbed in the evaporator and rejected in the condenser. Under a high pressure, the ammonia refrigerant condenses from a gas to a liquid in the condenser, giving up its latent heat in the process. The high-pressure liquid now passes through a refrigerant restrictor, where its pressure and temperature are reduced. This low-pressure liquid boils inside the evaporator to a gas, absorbing latent heat in the process and cooling off the water passing over the evaporator coil. So far, the compression and absorption systems are identical. The difference is in how the low and high pressures

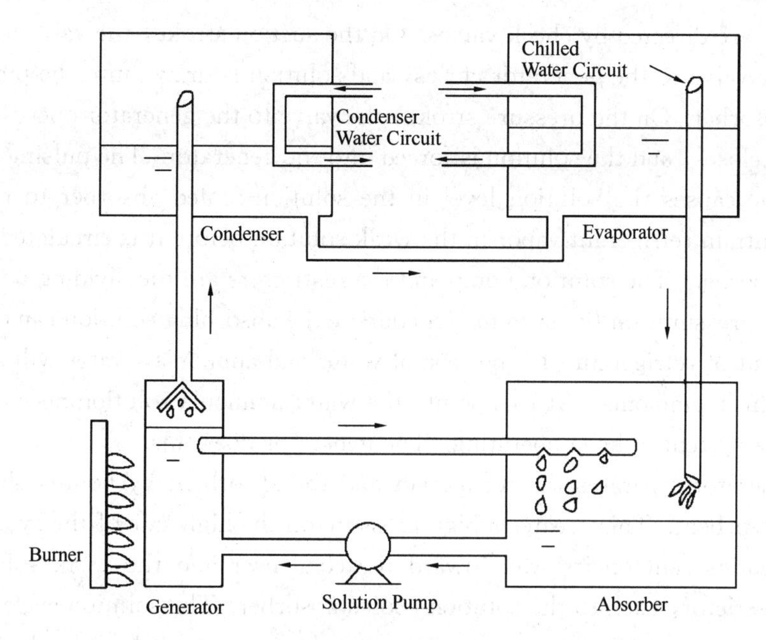

Figure 6-2 Simplified ammonia-water absorption cycle. The solution cycle is on the left; the refrigeration cycle is on the right.

are created.

In a compression system, the compressor provides the force to create both the high-side and low-side pressures. In anopabsorption system, the compressor is replaced by three components: the generator, the absorber, and the solution pump.

One component that absorption systems have that is not found on compression systems is the absorber. A second fluid that has an affinity for the refrigerant absorbs the refrigerant in the absorber. This has the effect of "sucking" the refrigerant out of the evaporator and creating a low refrigerant pressure in the evaporator. The absorption process replaces the suction stroke of the compressor. When the vapor refrigerant is absorbed into the solution, it condenses back into a liquid. This gives off a great deal of heat, normally called the heat of absorption. If the heat of absorption is not removed from the solution, the absorption process will stop. Some units actually have two absorbers: a solution-cooled absorber and an air-cooled absorber.

The solution-cooled absorber is where the refrigerant and weak solutionare first mixed. The solution-cooled absorber can not cool the solution adequately enough to remove all the heat of absorption; therefore, much of the refrigerant remains in a vapor state inside the solution. To remove the heat of absorption, the solution passes through the air-cooled absorber, where the heat of absorption is removed and the refrigerant vapor is condensed. A solution pump is used to pump the solution from the low pressure absorber to the high-pressure generator.

The solution pump is the closest thing to a compressor in concept. The solution pump used in many units is a diaphragm pump and pumps somewhat like a heart, in pulses of pressure that are directed by check valves. On the suction stroke, the valve to the absorber opens, the valve to the generator closes, and solution is drawn into the pump from the air-cooled absorber. On the pressure stroke, the valve to the generator opens, the valve to the absorber closes, and the solution is forced into the generator. The pulsing action of the solution pump causes the solution level in the solution-cooled absorber to rise and fall. This helps entrain refrigerant vapor in the weak solution before it is circulated through the air-cooled absorber. The solution pump and the restrictors are the dividing points between high and low pressures on the system. Of course, the absorbing solution can only absorb a limited amount of refrigerant. In the case of water and ammonia, water will absorb about half its weight in ammonia. At this point, the water-ammonia solution needs to be regenerated for the system to keep operating. The generator does this.

The generator separates the refrigerant and the absorbent by boiling the refrigerant out of the absorbent. This creates a high pressure on the high side of the system. The refrigerant vapor is sent on its way toward the condenser and the weak solution is sent through a restrictor, back to the solution-cooled absorber. The solution cycle is now complete. In addition to the components mentioned, there are a number of other components added to the system for increased efficiency.

Never use standard brass gauges to service an ammonia system! Ammonia attacks the brass. Ammonia systems require gauges made specifically for ammonia.

Evaporative Cooling Systems

Evaporative cooling reduces the air temperature by evaporating water. The heat required to evaporate the water comes from the air, reducing the temperature of the air. Air that has been cooled through direct evaporative cooling still has the same amount of heat, but its temperature is lower and its humidity is higher. This method works well in dry climates, such as in the southwestern United States. Evaporative coolers can drop the air temperature by as much as 20℉ if the entering air is dry. Evaporative cooling efficiency increases as the air temperature increases because the water evaporates more readily. Evaporative cooling does not work as well in humid climates, such as coastal areas or the southeastern United States. The high humidity in these areas reduces the effectiveness of evaporative cooling because the water evaporates slowly. The installation and operation costs for an evaporative cooler are much lower than conventional air conditioning using a refrigeration cycle. This makes evaporative cooling popular in large factories and plants, where traditional refrigerated air conditioning would be prohibitively expensive.

Outdoor misters are an increasingly popular form of direct evaporative cooling. Nozzles that spray outmicrodroplets of water cool the outside air through evaporation. Amusement parks now often use outdoor misters to provide outdoor cool spaces during the summer. Mist systems can also be applied to residential outdoor spaces such as patios.

A typical evaporative cooled air conditioner, also called a "swamp cooler." The unit consists of a water sump, a pump, wetted media to increase the surface area for evaporating water, and a blower to move air. The pump pumps the water in the sump at the bottom to the top of the media where it runs down the media. A float controls the water level in the sump by allowing makeup water to fill the sump when the water level drops. The blower moves outside air across the media and into the house. Water evaporates from the wetted media, reducing the temperature of the air. For this system to operate correctly, air in the house must be allowed to escape. This is normally done by opening the windows. Evaporative coolers are constantly replacing the air in the house with air from outside that has been cooled by evaporation. This type of evaporative cooling is called direct evaporative cooling because the water is directly exposed to the air entering the building.

Indirect evaporative cooling uses a heat exchanger and two airstreams. The heat exchanger keeps the airstreams separate while allowing one airstream to cool the other. Figure 6-3 shows an indirect evaporative cooler. The water evaporation takes place in the airstream that passes through the heat exchanger without entering the building. The airstream entering the building is cooled by the heat exchanger without adding any water to it. Indirect evaporative cooling can be used to precool air before it passes through a traditional compression-cycle conditioning system. This saves energy by reducing the amount of cooling that the compression system must provide as shown in Figure 6-4.

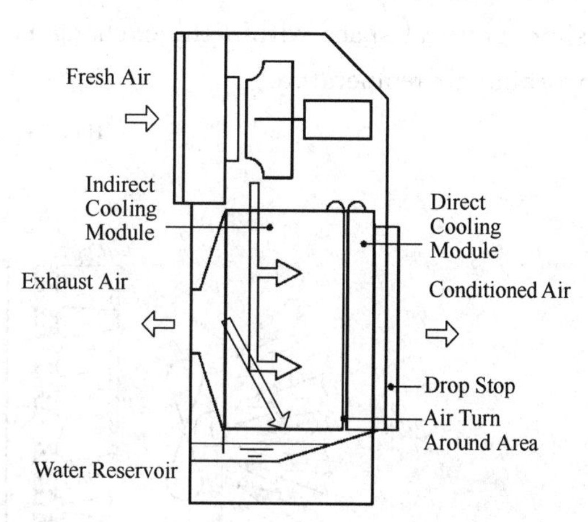

Figure 6-3 Indirect evaporative cooling process.

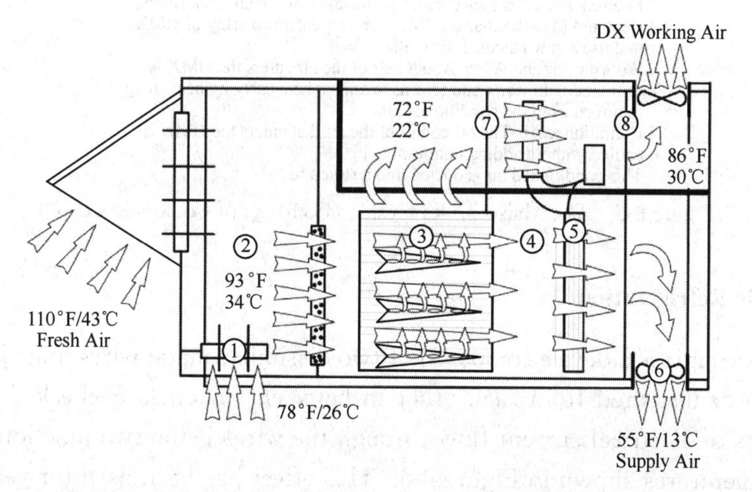

Figure 6-4 Indirect cooling can be used toprecool the air entering a traditional compression system. (Courtesy of Coolerado Corp.)

The effectiveness of evaporative cooling can be increased using multiple stages. One manufacturer, Coolerado, makes a twenty-stage indirect evaporative cooler using the Maisotsenko Cycle. The system uses a unique heat and mass exchanger (HMX), shown in Figure 6-5, that consists of several plates of a special plastic that is designed to wick water evenly on one side and transfer heat through the other side. The plates are stacked on each other, separated by channel guides. The channel guides divide the incoming air stream into product air and working air. The product air will be delivered into the house; the working air will be expelled to the outside. Product air is always separate from the working air so that the air entering the house does not pick up moisture. Air that is cooled by the first set of plates is used to cool the next set of plates. This process occurs multiple times in a short physical space within the exchanger, resulting in progressively colder product and working air temperatures.

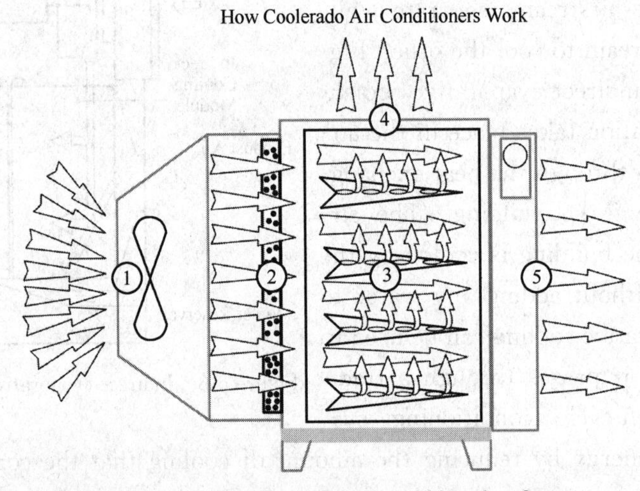

1. Fresh Air-Outside air is drawn into the air conditioner by a fan.
2. Filtered-The air is then cleaned by an array of 2 inch thick filters.
3. Heat and Mass Exchange(HMX)-The air enters an array of HMX that use a new patented technology.
4. Working Air and Water-About half of the air enters the HMX is saturated with water and returns to atmosphere carrying heat energy removed from the conditioned air.
5. Conditioned Air-The other half of the air that enters the HMX is cooled without adding humidity to it. This conditioned air is ducted into a structure.

Figure 6-5 The Maisotsenko Cycle. (Courtesy of Coolerado Corp.)

Thermoelectric Refrigeration

A thermocouple is a device composed of two dissimilar metal wires joined on both ends but separated or insulated from each other in between. Thomas Seebeck discovered that small amounts of electrical current flow through the wires if the two junction points are at different temperatures shown in Figure 6-6. This effect has been used for years as a flame safety device in gas-burning appliances. While studying the Seebeck effect in 1834, Jean Pelletier discovered that if a current is imposed on the thermocouple, one end will heat up and the other will cool off. Reversing the direction of the current will swap the hot and cold junctions. This is called the Pelletier effect shown in Figure 6-7. This characteristic is exploited in thermoelectric refrigeration.

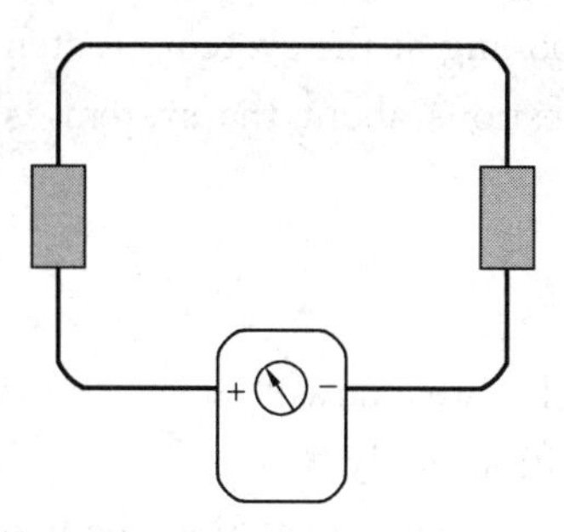

Figure 6-6 Temperature difference between the two junctions creates a DC current flow.

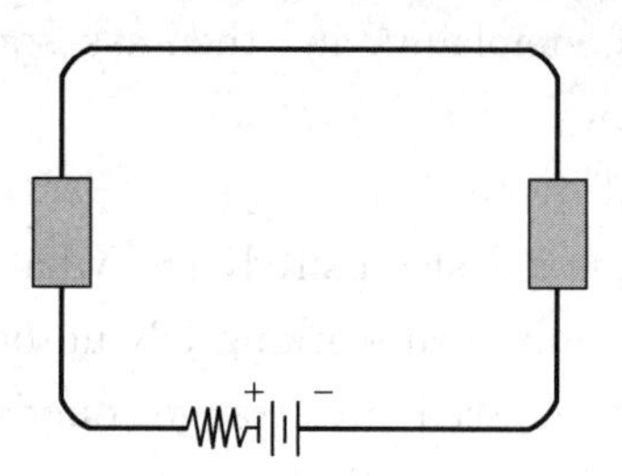

Figure 6-7 A DC current imposed across the two junctions creates a temperature difference.

Today's thermoelectric modules typically use semiconductors rather than metal. Figure 6-8 shows the construction of a typical thermoelectric module. These modules can be sandwiched between two heat sinks to produce a unit capable of moving heat. Figure 6-9 shows a complete thermoelectric cooling module, including the heat sinks. Thermoelectric refrigeration is not very energy efficient and is best suited for small loads. Thermoelectric cooling does provide a small, lightweight cooling system for areas that would be difficult to cool with a traditional refrigeration system. Thermoelectric cooling has been applied to electronic systems, spaceships, and picnic coolers.

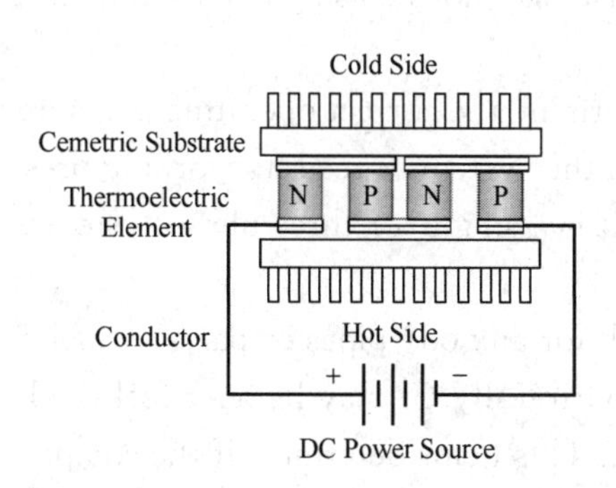

Figure 6-8 Modern thermoelectric modules use semiconductors to create the Pelletier effect.

Figure 6-9 Complete thermoelectric cooling module, including heat sinks.

6.3 Troubleshooting Refrigeration Systems

Patience is a requirement when troubleshooting refrigeration systems. It is important to take the time necessary to properly diagnose a system problem, as rushing to a quick diagnosis often leads to misdiagnosing the root cause of the problem. A good service practice is to spend some time speaking to the customer before looking at the system. Listen to the customer's complaints and then ask some pertinent questions about the system, such as the following:

- How old is the unit?
- When was it last repaired, and what was done?
- Has the unit been working OK up until the time of the breakdown?
- Have you noticed any strange sounds or erratic operation lately?

After speaking with the customer, it is time to gather some information about the system. Conduct a visual inspection of the system and look for any obvious system problems. This is another important part of the troubleshooting process. During the initial visual inspection try to determine the following:

- Unit model and serial numbers
- Required supply voltage and phase
- Current ratings of motors
- Full load amps for most motors
- Rated load amps (RLA) for compressor motors
- Type of metering device used
- Type of refrigerant in the system
- Condition of the evaporator and condenser coils
- Overall condition of the equipment

It may be necessary to remove panels on the equipment to see everything. A crucial part of the initial inspection is the examination of the evaporator and condenser coils. An iced-up evaporator coil and dirty condenser coil are two common problems that can easily be identified during a visual inspection. The cause may not be as apparent, but the symptom is easy to discover.

Identifying the refrigerant type helps to determine the correct operating pressure when it comes time to install a set of service gauges on the system to read its working pressures. Finding the correct supply voltage and whether it is a single- or three-phase system will also help when testing the circuit.

Another part of a visual inspection is to look for any oily pipes or parts, which is usually a good indication of a refrigerant leak. Also carefully (it may be hot) feel the head of the compressor to see if it is cold, warm, or hot. This helps determine if the compressor is the cause of the problem. For example, if the compressor is extremely hot and not running, it may be off due to an overload. You may want to start troubleshooting there to find out why this has happened.

Visually inspecting all the major components of a system takes a little extra time, but it is time well spent and is necessary to efficiently troubleshoot any refrigeration system.

Taking shortcuts on a job may save time but may cause a technician to work in an unsafe manner. This is not a wise tradeoff. Do not take any shortcuts that may cause you to work in an unsafe manner. It is simply not worth the few extra minutes you may have gained.

Electrical Troubleshooting

Electrical troubleshooting can be divided into two areas: (1) electrically diagnosing controls (switches, relays, and contactors) and (2) electrically diagnosing loads (coils, fan motors, and compressors). There are three major tools commonly used to troubleshoot the various electrical components of a refrigeration system: (1) voltmeter, (2) ammeter, and (3) ohmmeter. Each measures a different electrical characteristic in a circuit and is needed to properly diagnose an electrical fault within a system. Many technicians will use a multimeter to measure all of these electrical characteristics.

When replacing or repairing any electrical components, always verify that the voltage source is truly disconnected from the circuit. Test the circuit for the presence of voltage with some type of voltmeter or voltage indicator. Do not solely rely on the electrical disconnect to ensure the voltage is disengaged. Always verify this yourself.

A voltmeter measures electrical potential (voltage) at a switch or a load (such as a motor or the coil of a contactor). For a load to operate properly, the correct voltage must be applied. Most units will have the required applied voltage stamped on a clearly marked nameplate. If a load is not energized, one of the first steps in the troubleshooting process is to determine if the proper voltage is applied to th enonenergized load. For most loads, the applied voltage must be within 10 percent of the nameplate voltage. If no voltage or the incorrect voltage (outside the 10 percent tolerance) is present at a load, then the problem lies in the voltage supply, controls, or wiring leading to that load. When checking the applied voltage to a load, always measure the voltage at the load and not at an electrical junction before the load. There may be an issue with the wiring leading to the load, so it is more accurate to measure the applied voltage directly there.

A voltmeter can also be used to check the operation of the various electrical controls in a circuit. Electrical controls fail in one of two ways: they fail to close or they fail to open. A voltmeter can be used to determine whether a control is electrically open or closed by measuring the line voltage across the control's contacts. A measured voltage indicates that the contacts are electrically open. If the contacts should be closed, then the circuit problem lies in the control.

When working on electrical circuits, safety should always be first on your mind. Electrical circuits should always be de-energized before repairing or replacing any electrical component or wiring. De-energize at the system's disconnect and then follow standard lockout/ tag-out procedures.

A zero measured voltage would seem to indicate the control is closed. However, this is not always true. The absence of a line voltage across a contact does not always indicate the contact is closed. If two or more controls wired in series are electrically open in a circuit, then there will not be a line voltage drop across their contacts. Therefore, this method of electrically troubleshooting a control is not always reliable. A more reliable method of electrically troubleshooting controls is a procedure referred to by many technicians as hopscotching. This allows a technician to easily isolate an open electrical component. First, attach one probe of a voltmeter to a common point in the circuit as shown in Figure 6-10. The blue probe is attached to the L2 leg of the circuit and will stay attached at this point during the process. This is an electrically common point for all the loads in the circuit. Now the red probe can be moved around the circuit to help determine which switch is electrically open or closed.

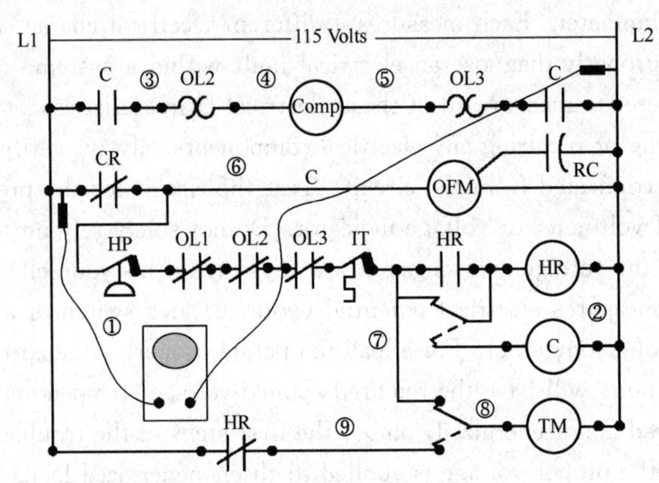

Figure 6-10 Hopscotching an electrical circuit.

Here is an example of the process. As shown in Figure 6-10, the red probe of the voltmeter is initially connected before the CR contact and should read line voltage, which is 115 V in our example. When the red probe is moved to point 6 on the diagram, 115 V should again be read on the voltmeter if the CR contact is electrically closed. If the voltmeter reads 0 V, the CR contact is electrically open. This process can be repeated for all of the switches in the circuit. If line voltage is read before and after the switch, the switch is electrically closed. If line voltage is read before but not after the switch, the switch is electrically open.

Mechanical Refrigeration Troubleshooting

Troubleshooting the mechanical side of a system involves analyzing or investigating the refrigerant flow either throughout the entire system or through any of the components. This requires being able to determine the refrigerant's pressure and temperature at various locations throughout a system.

When analyzing the entire system, the actual refrigerant conditions at various locations in a system are compared to the design conditions of a properly operating system. This will normally allow a technician to determine a system's defect. For example, if a system is discovered to be operating with both lower than normal suction and discharge pressures, and the superheat value of the refrigerant leaving the evaporator is higher than normal with a lower than normal subcooling value leaving the condenser, a technician can determine that the system has a low refrigerant charge.

Many times a technician will need to analyze the process through a component such as the compressor, condenser, or evaporator. This requires an understanding of the component and how it is designed to function in a system. For example, a refrigeration compressor is designed to pressurize a low-pressure refrigerant vapor to a higher-pressure refrigerant vapor. If a compressor is electrically operating but fails to sufficiently increase the pressure of the refrigerant from its inlet to its outlet, the compressor is mechanically defective, and more than likely the compressor will need to be replaced.

In general, a technician can analyze the operation of the evaporator as well. If the superheat value of the refrigerant leaving an evaporator is higher than normal, the evaporator is starved for refrigerant. If the superheat value of the refrigerant is lower than normal, the evaporator is flooded with refrigerant. Finding the reason why the evaporator is either being starved or flooded, however, will require looking at the entire system and analyzing other conditions.

When analyzing the condition of the refrigerant throughout a system or through a system component, a technician must allow the pressures and temperatures of the refrigerant to stabilize upon restarting if the system has been off. When troubleshooting, it is a good service practice to let a system run for about 10 minutes before recording any of the system's pressures or temperatures.

Troubleshooting a Defective Compressor

Replacing a compressor on a refrigeration system is never an easy or inexpensive task. If a compressor is found to be defective, every effort should be made to verify that a correct diagnosis is made. Sometimes compressors that have been changed out in the field are later found to be operational.

Compressor problems can be divided into two groups: mechanical and electrical. Mechanical defects are problems that affect the operation inside the compressor, such as blown valve plates, a broken crankshaft, or worn pistons. These defects will either cause a compressor not to pump any refrigerant or pump to below its rated capacity.

It is generally easy to determine if a compressor is not pumping adequately. At start-up, verify that the compressor is electrically energized, then measure its amperage draw. It will be lower than normal. Next, monitor the compressor's suction and discharge pressures. Neither the suction nor the discharge pressure will change dramatically. There may be a very slight change in pressure, but the suction pressure will stay very high and the discharge will stay low.

To determine whether a compressor is pumping to its rated capacity is not quite as simple. A technician must measure the compressor's actual amperage draw and compare this to the amperage draw as stated by the manufacturer. The amperage should be within ±5 percent of the manufacturer's stated value. If it is drawing less than 5 percent of its rated amperage draw, it is not pumping to its rated capacity.

Compressors can fail to start as a result of either an electrical defect or a mechanical defect. Mechanically, one or more of the pistons within the compressor can become locked and fail to move. The compressor will attempt to start, but since the pistons will not move, the compressor draws high amperage, causing it to shut down on overload.

There are three major electrical defects that will cause a compressor not to operate. One possibility is the motor windings of the compressor are open, shorted, or grounded. Another possibility is that the starting relay or capacitor is defective. The third possibility is that the incorrect voltage is applied to a compressor.

To check the condition of the compressor's motor windings, a technician will need to measure the resistance of the windings using a standard ohmmeter. For single-phase compressors, measure the resistance value of both the run and start windings and measure the resistance of each winding to ground. For three-phase compressors, each of the three windings will need to be checked. If the correct resistance value is measured across its windings, as shown in Figure 88-6 , and an OL is measured from the windings to ground, then the windings are satisfactory.

The voltage applied to a three-phase motor must bebalanced, meaning the applied voltage to each leg must be relatively the same. The applied voltage must not deviate more than 2 percent from the average supplied voltage. A voltage imbalance of greater than 2 percent will cause the windings within a motor to generate heat beyond safe levels, leading to premature motor failure.

There is a common scenario in which a compressor may appear to be defective when in fact it is not. A compressor with an internal overload that has overheated will shut down and will not restart until the compressor has cooled. If a technician performs a resistance check on the compressor, the first step will be to measure the resistance from common to run and common to start, which will be OL. The technician may interpret this to be an open winding in the compressor. However, if the compressor is allowed to cool down and the internal overload to reset, the technician may find that the compressor will start normally and will not need to be replaced. The reason why the compressor overheated will need to be identified and resolved.

A compressor with a defective run capacitor will draw higher than normal amperage and cycle off on its overload or not start at all. If a compressor incorporates a start relay and a start capacitor and either is defective, the compressor will most likely not start.

Incorrect voltage applied to a compressor may cause it to run for a brief time then cycle off on its overload or not start at all. A service technician must first verify the correct voltage for the compressor and then measure to see what actual voltage is applied. Most compressors are rated with a tolerance of 10 percent. If the applied voltage is outside these limits or those stated by the manufacturer, it must be corrected before the compressor can be properly diagnosed.

Troubleshooting Evaporator Problems

An evaporator is a heat exchanger used to absorb heat from the refrigerated product. Heat is transferred from the product through the tubing and fins of the evaporator to the refrigerant. If the evaporator's surface becomes insulated as the result of either being dirty or covered with ice, the amount of heat transfer will be dramatically reduced.

Ice buildup can occur when the heat load on the evaporator is reduced. As a result, the temperature of the evaporator drops and an excessive amount of frost and ice accumulates on its surface. For example, a blower-type evaporator needs the correct amount of airflow across its surface. If this airflow is restricted as a result of one or more defective fan motors, the surface of the evaporator will develop an excessive amount of ice and the system will not be able to absorb enough heat from the product to maintain the desired case temperature.

With an evaporator temperature below 32℉, frost will naturally develop on the surface of the evaporator even if the correct heat load is applied. The design of the refrigeration system must incorporate some means of defrosting the evaporator on a regular basis to prevent an excessive amount of frost from developing on the surface of the evaporator. Medium-temperature refrigeration systems will use the case air to defrost the evaporator. When the refrigeration system shuts down, the evaporator fan continues to operate and will continue to draw case air across the evaporator to melt any frost that has accumulated. However, with low-temperature systems the case temperature is well below 32℉ and therefore the case air cannot be used to defrost the evaporator. Low-temperature systems require some type of supplemental heat to defrost the evaporator surface.

A frozen evaporator coil is a common problem found when troubleshooting freezers. Although there are several possible causes for this problem, one common cause is the defrost system. The system is not properly defrosting the evaporator's coil on a regular basis. To effectively troubleshoot this problem, a technician must understand the design and operation of the defrost systems typically used.

Defrost system problems fall into two categories. The system may be over-defrosted or under-defrosted. Overdefrosting is when the defrost heater stays energized too long, causing the box temperature to rise too high during defrost. The product may begin to melt and then refreeze. This problem can be identified by monitoring the box temperature during defrost or by examining the product in the case. Ice crystals forming on frozen product may be an indication of melting and refreezing. Over-defrosting is normally caused by a malfunction in the defrost cycle termination. On time defrost systems, either the defrost time is set too long or the time clock is defective. On time-pressure systems, either the terminating pressure control is set too high, the pressure control is defective, or the solenoid coil in the defrost timer is incorrectly wired or defective. On time-temperature systems, either the temperature control is set too high, the control is defective, or the solenoid in the defrost timer is defective.

Systems that are under-defrosting will result in a frozen evaporator. The frost that normally develops on the evaporator coil will continue to build until the entire surface of the evaporator is iced over. This can be caused by a defective or incorrectly set time clock that does not initiate a defrost cycle. It can also be caused by an open heater element or a defective defrost termination switch that continually terminates defrost on each initiation of a new cycle.

When an evaporator ices over, a technician will need to defrost the coil. Sometimes this can be done by manually initiating a defrost cycle. This can be done on a mechanical defrost timer by rotating the inner knob counter-clockwise. Many times, however, this cannot be done. The technician will have to manually defrost the coil. Extreme care must be taken when doing this. Do not use any sharp objects to chip away ice from the coil—this could easily puncture the coil and cause a refrigerant leak to develop in the evaporator. Using water is the best method but is not always practical, since the water will need to be drained away. If water cannot be used easily, a heat gun usually works well. Defrosting a coil manually is time-consuming. Do not rush this procedure, as making a careless mistake can be costly.

Unit 7 Air Conditioning Systems

In 1911, Mr. Carrier, who is called the"father of air conditioning," presented his "Rational Psychrometric Formulae" to the American Society of Mechanical Engineers. Today, the formula is the basis in all fundamental psychrometric calculations for the air conditioning industry. Though Willis Carrier did not invent the first air conditioning system, his cooling and humidity control system and psychrometric calculations started the science of modern psychrometrics and air conditioning. Air "cooling" was only part of the answer. The big problem was how to regulate indoor humidity. Carrier's air conditioning invention addressed both issues and has made many of today's products and technologies possible. In 1900s, many industries began to flourish with the new ability to control the indoor environmental temperature and humidity levels in both occupied and manufacturing areas. As already mentioned, today, air conditioning is required in many residential homes, office building, and most industries and especially in ones that need highly controllable environments, such as clean environment rooms for medical or scientific, product testing, and sophisticated computer and electronic component manufacturing.

7.1 Residential Air Conditioning

Residential air conditioning generally includes single family, multifamily, and low-rise multifamily private household residences. The air-conditioning type is typically central forced air, central hydronic, or zoned. The capacities of residential installations typically range from less than 1 ton up to about 5 tons.

Unitary Systems

Residential air-conditioning systems are often made up of unitary air-conditioning components. These are factory built and tested systems, complete as much as possible, with piping, controls, wiring, and refrigerant. Self-contained package units are usually simple to install, requiring only service connections and in some cases ductwork for field applications. The simplest of these is the window-mounted room air conditioner.

Unitary air-conditioning equipment consists of one or more factory-made assemblies, which normally include an evaporator or cooling coil, a compressor and condenser combination, and possibly a heating unit. When the air conditioner is connected to a remote condensing unit, such as in residential applications, the system is often referred to as a split system. A split-system installation will require more fieldwork for the technician as compared to the installation of a simple package unit. The sizes of unitary equipment range from small fractional-tonnage room coolers to large packaged rooftop units in the 100-ton category. Split-system equipment up to 5 tons in capacity may be classified as either commercial or residential. There is a wide range of applications in both markets using the same product; however, above 5 tons the application becomes distinctly commercial, and product designs use different components.

Room Air Conditioners

Room air conditioners were primarily developed to provide a simplified means of adding air conditioning to an existing room. These units are considered semiportable in that they can easily be moved from one room to another or from one building to another. They provide cooling, dehumidifying, filtering, and ventilation, and some units provide supplementary heating.

In numbers sold, room air conditioners such as the one shown in Figure 7-1 outsell all other types of unitary equipment. They are relatively low in cost, easy to install, and can be used in almost any type of structure, as shown in Figure 7-2 .

Figure 7-1 Window air-conditioning unit for cooling.

Figure 7-2 Window air-conditioning unit on the Harry ElkinsWidner building at Harvard University.

The disadvantage of room air conditioners is that they may either block part of the window area and prevent the window from being opened or require a special hole through the wall. Some people object to the operating noise that they produce close to the occupants. They are best used to condition a single room, but the spillover can supply some conditioning to adjacent areas.

Construction and Installation of Room Air Conditioners

A common problem with window air-conditioning units is the nuisance trips of the residence circuit breakers. Window units are often used in older homes where the electrical service is not adequate for the air-conditioner load. It may be possible to rewire the residence for a single circuit to the air conditioner, or it may be best to downsize the system to one that will not trip the circuit breaker. Downsizing the unit may provide less than desired levels of cooling on extremely hot days. However, the constant shutting down of the oversized unit electrical service can result in the same lack of cooling. In addition, it may be possible to use more than one window unit in a room if the electrical outlets in that room are serviced by more than one circuit breaker.

There are basically two parts to the unit. One section goes inside the room, where the evaporator fan draws in room air through the filter and cooling coil, delivering conditioned air back into the room. The other section extends outside the room, where the condenser fan forces outside air through the condenser, exhausting the heat absorbed by the evaporator. One motor operates both the indoor blower and outdoor fan. The motor shaft extends through the separating partition and drives both fans. Condensate from the evaporator coil flows into the drain pan, which extends below the condenser fan. The condenser fan tip dips into condensate, splashing it onto the hot condenser, where it evaporates and is blown into the outside air.

The window-mounted units are supplied with a kit of parts for installation. Sill brackets, window mounting strips, and sealing strips are set in place for installations in double-hung windows. Side curtains fold out to fill up the extra window space. A sponge rubber seal is provided for the opening where the sash overlaps, and a sash bracket is installed to lock the lower sash in place. Room air conditioners are also available in a vertical configuration for mounting in sliding window openings. These vertical conditioners are held in place by the sides as opposed to the conventional-shaped room air conditioners that are secured in place on their tops and bottoms.

Be sure to plan the installation, even if you feel it will be simple! Very often there will be furniture in the way of the window. Ask for permission to move the furniture or have the customer clear the space. Drilling and mounting the unit will be somewhat messy. Warn the customer of this ahead of time. Place a drop cloth on the floor and then always remember to clean up any mess when you are finished with the job.

There are also a number of safety concerns that must be addressed. Some units are quite heavy and can easily be dropped from a high-story window during installation. Many units have sharp corners and are very slippery, so a good pair of gloves will help. Grabbing onto the tubing for a better hold will ruin the new unit before it is even installed. Don't drill through the tubing when installing the cabinet with the chassis inside. Make sure that the supporting bracket is installed properly and in place before aligning the unit. If the unit is too heavy, then get some additional help. When lifting the unit with another person, keep it level. The taller person will tend to tip the unit and the drain pan of water will spill all over you and the customer's floor.

EER AND SEER Ratings

The typical size range for window air conditioning units is from 5,000 to 20,000 BTU/hr. These ratings are based on inside air at 80°F db temperature 67°F wb temperature, and outside air at 95°F db temperature. Smaller units normally operate on 115 V, while larger models operate on 230/208 V.

Equipment is generally rated for efficiency by the manufacturer to help determine which system is most suitable for a particular application. This energy efficiency ratio (*EER*) is calculated by taking the output in BTU/hr and dividing it by the input in watts. The higher the EER rating, the better the efficiency.

The seasonal energy efficiency ratio (SEER) takes into account the cycling of the equipment on and off and is generally considered to be a better indicator of actual equipment efficiency. The Air Conditioning, Heating and Refrigeration Institute (AHRI) rates and publishes these ratings.

Console Through-The-Wall Conditioners

A console through-the-wall conditioner is a type of room cooler that is designed for permanent installation. It was developed to provide individual room conditioning for hotels, motels, offices, and low-rise multifamily private household residences where it is impractical or uneconomical to install a central plant system. An opening needs to be made in the outside wall adjacent to the unit for condenser air and ventilation. These units are also known as packaged terminal air conditioners (PTAC).

PTACs are most commonly used in hotels, nursing homes, apartment complexes, and other such areas where the area being cooled is limited. PTAC units may be straight air conditioning, heat pumps, heat pumps with electric resistance supplemental heating, or straight air conditioning with electric resistance heating. All of these units are similar in appearance, and most fit in a standard wall sleeve. These units are accessible from inside the occupied space, and in most cases the condensate is simply allowed to drain outside the building. Some cities and municipalities have ordinances controlling condensate from PTAC units, but many do not.

PTAC sizes range from 6,800 to 14,900 BTU/hr at standard rating conditions. The quantity of outside air that the units can admit ranges from 40 to 55 CFM depending on the size of the unit. Units are available for 115 and 208/230 V, singlephase, AC power. Electric heat capacity from 1.5 to 5.0 kW can be installed in any size unit. Power-receptacle configurations depend on the amperage drawn.

PTAC units are efficient, quiet, and easy to install. Temperature efficiency is usually stated in terms of EER. The EER is equal to the cooling output in BTU/hr divided by the power input in watts under standard rating conditions. Standard rating conditions set up by AHRI are based on 80°F db, 67°F wb for the indoor entering air, and 95°F db for the outdoor ambient air. For example, a unit with an output of 6,600 BTU/hr and 660 W input, under standard conditions, would have an EER of 10 (6,600/660), which is considered a good rating.

Stand-Alone Air-Conditioning Units

A stand-alone air-conditioning unit for residential use requires an air handler outfitted with an evaporator coil, metering device, drain pan, transformer, blower assembly, and housing. The air handler is connected to its own duct system. There are three configurations of air handlers: vertical upflow, vertical downflow, and horizontal.

The major disadvantages of a stand-alone air-conditioning unit are the type and location of the condenser. The stand-alone will be conveniently located inside the conditioned building. Therefore it is impractical to use an air-cooled condenser unless the heat can be rejected to the outdoors in some convenient fashion. Most stand-alone air-conditioning units will have a water-cooled condenser. This will require a continuous source of cooling water or some fashion of cooling water loop.

Residential Split Systems

For many air-conditioning systems, it is not practical to place all components in a single package, particularly those that involve the use of air-cooled condensers. The air-cooled condenser must have access to outside air, so it is best placed outside. For this reason, split systems have been developed with the inside unit consisting of a fan coil unit, with or without heating; an outside mounted air-cooled condensing unit; and connecting refrigerant piping between the two.

The final stage in manufacturing for split condensing systems is at the residence. Manufacturers have no control over the individuals who perform this last vital step in the production of their equipment. Manufacturers can't even insist that a matching system is installed. This final stage in manufacturing has a greater influence on the performance of a system than anything that the manufacturer can do. If the system is not installed with the correct size air duct systems, electrical wiring, and many other factors, the result can be the system providing poor performance or a reduction of its operational life. It is therefore very important that all of the manufacturer's specifications and guidelines be followed during installation so that your customer's system can be its most efficient and have the longest life possible.

A typical split outdoor air-cooled condensing unit for residential settings consists of a compressor, condenser coil, a condenser fan, and the necessary electrical control box assembly. On residential condensing units, a fully hermetic compressor is used. The condenser coil is a fin-and-tube arrangement that varies in design from manufacturer to manufacturer. A large surface area is desirable, and many units offer almost a complete wraparound coil to gain maximum coil surface area. The coil tube depth is limited to reduce resistance to airflow.

Condenser fan motors turn at slower speeds than indoor blower motors. Not all condenser fan motors spin at the same rpm. Some operate around 1,000 rpm, while others may operate around 800 rpm. It is important when changing out a condenser fan that one of the exact same rpm be used. The fan blades are designed to move air at a certain speed. If the fan blade is spinning faster than it was designed for, it will not operate as efficiently. Increasing the rpm does not mean that you are going to increase the airflow. In addition, the height that the fan extends into the fan shroud is important. If the fan blade is too low or too high, it will affect the airflow from the condenser. Mark the height of the blade before removing it to be sure it is put back in at the same height.

The condenser fan varies in design but is usually a propeller-type fan, which can move large volumes of air through clean coils, which in turn offer low resistance. Airflow direction is a function of the cabinet and coil arrangement, and there is no one best arrangement. Most units, however, use draw-through operation over the condenser coil. Outlet air can have an effect on surrounding plant life. Top discharge is the most common arrangement. Fan motors are sealed or covered with rain shields. Fan blades are shielded with a grille for the protection of hands and fingers.

For evaporator coils, coil cases or cabinets are insulated to prevent sweating, and all have pans for collecting condensate water runoff. Plastic pipe may be used to connect condensate drain water to the nearest drain. If no nearby drain is available, a small condensate pump may be installed to pump the water to a drain or to an outdoor disposal arrangement.

Heat Pump

Heat pumps are commonly used for split systems. This is because heat can either be rejected or absorbed by the outside unit. On hot days in the summer when air conditioning is required, the unit will operate like a normal split air-conditioning system. The outside unit will operate as a condenser rejecting heat, and the inside unit will operate as an evaporator to cool the inside air. On colder days in the winter, the system will reverse and the outside unit will absorb heat from the outside air. This heat is transferred to the indoor unit, which now operates as the condenser and heats the inside air. This type of air-to-air heat pump is very common in climates where the outside temperatures do not fall very far below freezing in the winter. When outside temperatures do fall below the normal operating range for the unit, an electrical resistance heater will turn on. The cost for supplementing with this type of electric heat, however, is often far more expensive as compared to operating the normal heat pump cycle. Even so, heat pumps are used in colder climates, but they are arranged differently.

An outdoor unit for an air-to-air heat pump can operate as either a condenser or an evaporator. The indoor unit is suspended from the roof in the attic by chain hangers to save space. The indoor unit has a drain pan located underneath it to catch any condensate to prevent leaks coming through the ceiling below. This indoor unit also has a water loop that is augmented by solar heating panels located outside, on the roof of the building.

Mini-Split

Mini-split systems are similar to conventional split systems except they are smaller. They are typically used to cool or heat one space and not the whole building. The major advantage is that air ducting is not necessary. In some ways these are similar to window air-conditioning units and PTACs, but they do not require a window and they are more versatile. For example, mini-split indoor units can be placed anywhere in the home and multiple indoor units can be connected to a single outdoor unit. Mini-splits are available for cooling only or as heat pumps. This is possible because there is an outside unit and an inside unit. For installation, all that is required is a drilled hole through an outside wall for the piping and wiring connections.

7.2 Central Air Conditioning Systems

Historical Development

As part of a heating system using fans and coils, the first rudimentary ice system in the United States, designed by McKin, Mead, and White, was installed in New York City's Madison Square Garden in 1880. The system delivered air at openings under the seats. In the 1890s, a leading consulting engineer in New York City, Alfred R. Wolf, used ice at the outside air intake of the heating and ventilating system in Carnegie Hall. Another central ice system in the 1890s was installed in the Auditorium Hotel in Chicago by Buffalo Forge Company of Buffalo, New York. Early central heating and ventilating systems used steam-engine-driven fans. The mixture of outdoor air and return air was discharged into a chamber. In the top part of the chamber, pipe coils heat the mixture with steam. In the bottom part is a bypass passage with damper to mix conditioned air and bypass air according to the requirements.

Air conditioning was first systematically developed by Willis H. Carrier, who is recognized as the father of air conditioning. In 1902, Carrier discovered the relationship between temperature and humidity and how to control them. In 1904, he developed the air washer, a chamber installed with several banks of water sprays for air humidification and cleaning. His method of temperature and humidity regulation, achieved by controlling the dew point of supply air, is still used in many industrial applications, such as lithographic printing plants and textile mills.

Perhaps the first air-conditioned office was the Larkin Administration Building, designed by Frank L. Wright and completed in 1906. Ducts handled air that was drawn in and exhausted at roof level. Wright specified a refrigeration plant which distributed 10℃ cooling water to air-cooling coils in air-handling systems.

The U. S. Capitol was air-conditioned by 1929. Conditioned air was supplied from overhead diffusers to maintain a temperature of 75°F (23. 9℃) and a relative humidity of 40 percent during summer, and 80°F (26. 7℃) and 50 percent during winter. The volume of supply air was controlled by a pressure regulator to prevent cold drafts in the occupied zone.

Perhaps the first fully air conditioned office building was the Milan Building in San Antonio, Texas, which was designed by George Willis in 1928. This air conditioning system consisted of one centralized plant to serve the lower floors and many small units to serve the top office floors.

In 1937, Carrier developed the conduit induction system formultiroom buildings, in which recirculation of space air is induced through a heating/cooling coil by a high-velocity discharging airstream. This system supplies only a limited amount of outdoor air for the occupants.

The variable-air-volume (VAV) systems reduce the volume flow rate of supply air at reduced loads instead of varying the supply air temperature as in constant-volume systems. These systems were introduced in the early 1950s and gained wide acceptance after the energy crisis of 1973 as a result of their lower energy consumption in comparison with constant-volume systems. With many variations, VAV systems are in common use for new high-rise office buildings in the United States today.

Because of the rapid development of space technology after the 1960s, air conditioning systems for clean rooms were developed into sophisticated arrangements with extremely effective air filters. Central air conditioning systems always will provide a more precisely controlled, healthy, and safe indoor environment for high-rise buildings, large commercial complexes, and precision-manufacturing areas.

Fundamentals

Central air conditioning systems are also called central hydronic air conditioning systems. In a central hydronic air conditioning system, air is cooled or heated by coils filled with chilled or hot water distributed from a central cooling or heating plant. It is mostly applied to large-area buildings with many zones of conditioned space or to separate buildings.

Water has a far greater heat capacity than air. A comparison of these two media for carrying heat energy at 68°F (20℃) is shown inTable 7-1.

The heat capacity per cubic foot (meter) of water is 3466 times greater than that of air. Transporting heating and cooling energy from a central plant to remote air-handling units in fan rooms is far more efficient using water than conditioned air in a large air conditioning project. However, an additional water system lowers the evaporating temperature of the refrigerating system and makes a small- or medium-size project more complicated and expensive.

A central hydronic system of a high-rise office building, the NBC Tower in Chicago, is illustrated in Figure 7-3. A central hydronic air conditioning system consists of an air system, a water system, a central heating /cooling plant, and a control system.

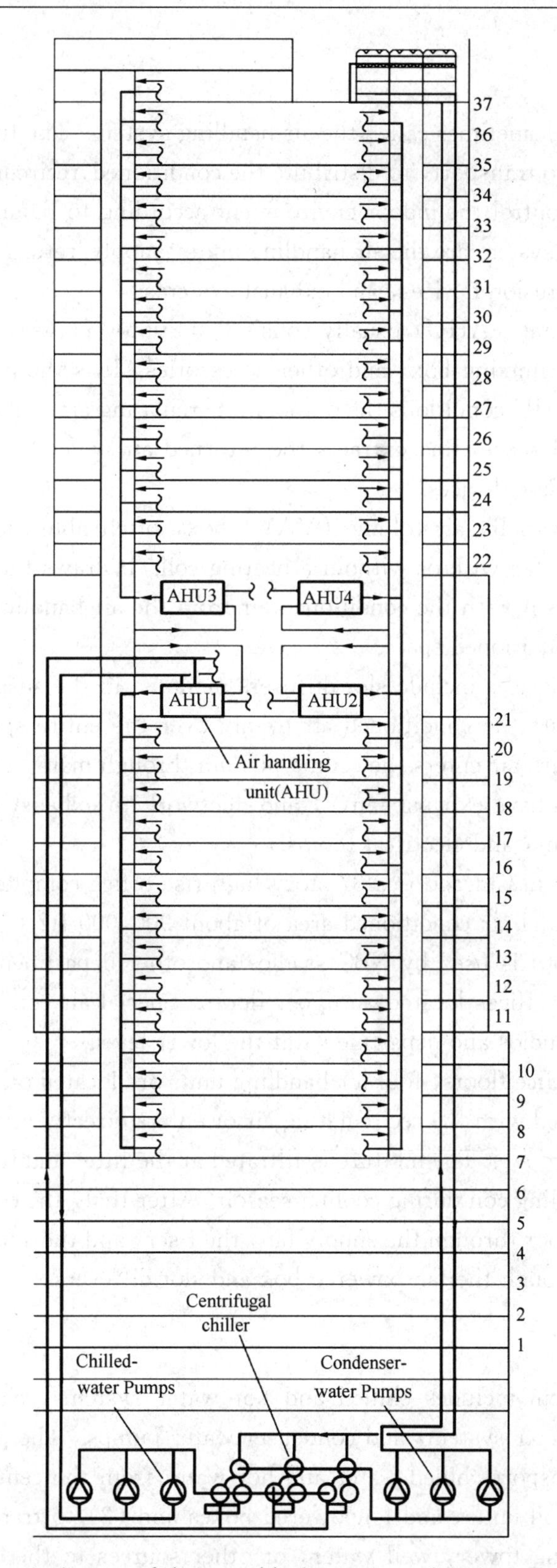

Figure 7-3 Schematic diagram of the central hydronic air conditioning system in NBC Tower.

Air System

An air system is sometimes called the air-handling system. The function of an air system is to condition, to transport, to distribute the conditioned, recirculating, outdoor, and exhaust air, and to control the indoor environment according to requirements. The major components of an air system are the air-handling units, supply/return ductwork, fan-powered boxes, space diffusion devices, and exhaust systems.

An air-handling unit (AHU) usually consists of supply fan(s), filter(s), a cooling coil, a heating coil, a mixing box, and other accessories. It is the primary equipment of the air system. An AHU conditions the outdoor/recirculating air, supplies the conditioned air to the conditioned space, and extracts the returned air from the space through ductwork and space diffusion devices.

A fan-powered variable-air-volume (VAV) box, often abbreviated as fan-powered box, employs a small fan with or without a heating coil. It draws the return air from the ceiling plenum, mixes it with the conditioned air from the air-handling unit, and supplies the mixture to the conditioned space.

Space diffusion devices include slot diffusers mounted in the suspended ceiling; their purpose is to distribute the conditioned air evenly over the entire space according to requirements. The return air enters the ceiling plenum through many scattered return slots.

Exhaust systems have exhaust fan(s) and ductwork to exhaust air from the lavatories, mechanical rooms, and electrical rooms.

The NBC Tower in Chicago is a 37-story high-rise office complex constructed in the late 1980s. It has a total air conditioned area of about 900,000 ft2 (83,600m^2). Of this, 256,840 ft^2(23,870m^2) is used by NBC studios and other departments, and 626,670 ft^2 (58,240m^2) is rental offices located on upper floors. Special air conditioning systems are employed for NBC studios and departments at the lower level.

For the rental office floors, four air-handling units are located on the 21st floor. Outdoor air either is mixed with the recirculating air or enters directly into the air-handling unit as shown in Figure 7-4. The mixture is filtrated at the filter and is then cooled and dehumidified at the cooling coil during cooling season. After that, the conditioned air is supplied to the typical floor through the supply fan, the riser, and the supply duct; and to the conditioned space through the fan-powered box and slot diffusers.

Water System

The water system includes chilled and hot water systems, chilled and hot water pumps, condenser water system, and condenser water pumps. The purpose of the water system is (1) to transport chilled water and hot water from the central plant to the air-handling units, fan-coil units, and fanpowered boxes and (2) to transport the condenser water from the cooling tower, well water, or other sources to the condenser inside the central plant.

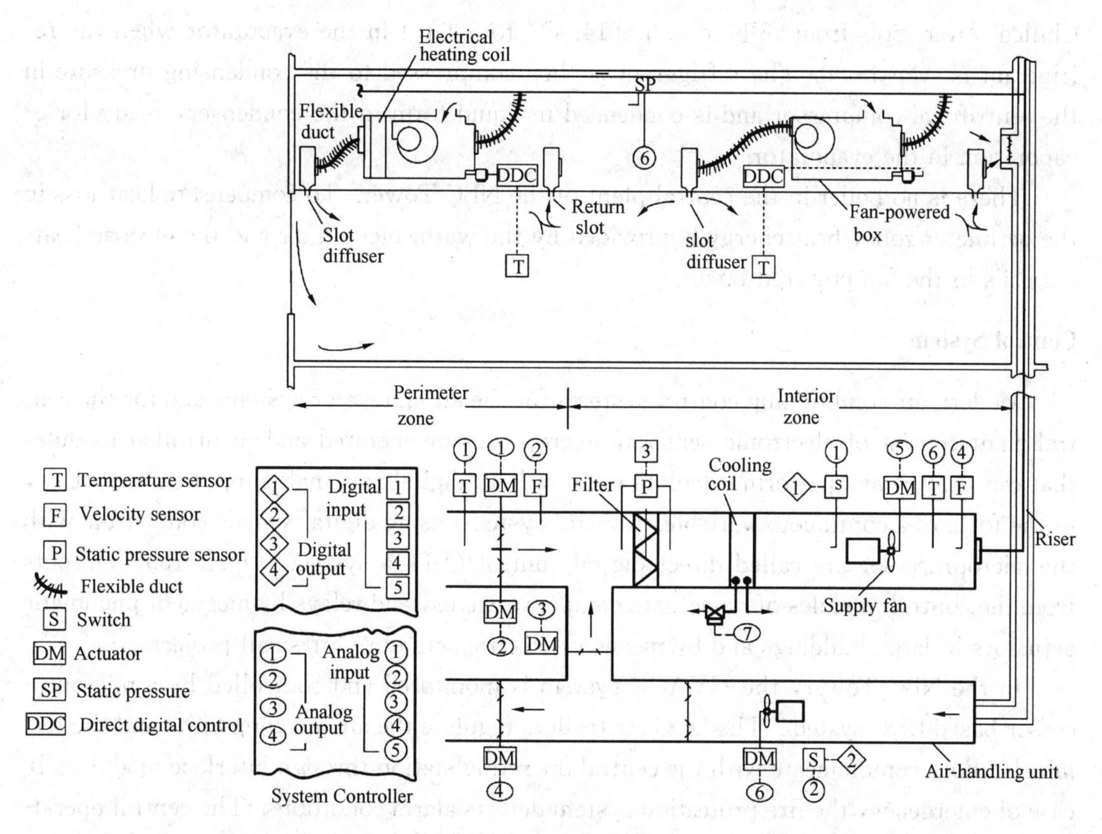

Figure 7-4 Schematic drawing of air system for a typical floor of offices in the NBC Tower.

In Figure 7-3 and Figure 7-4, the chilled water is cooled in three centrifugal chillers and then is distributed to the cooling coils of various air-handling units located on the 21st floor. The temperature of the chilled water leaving the coil increases after absorbing heat from the airstream flowing over the coil. Chilled water is then returned to the centrifugal chillers for recooling through the chilled water pumps.

After the condenser water has been cooled in the cooling tower, it flows back to the condenser of the centrifugal chillers on lower level 3. The temperature of the condenser water again rises owing to the absorption of the condensing heat from the refrigerant in the condenser. After that, the condenser water is pumped to the cooling towers by the condenser water pumps.

Central Plant

The refrigeration system in a central plant is usually in the form of a chiller package. Chiller packages cool the chilled water and act as a cold source in the central hydronic system. The boiler plant, consisting of boilers and accessories, is the heat source of the heating system. Either hot water is heated or steam is generated in the boilers.

In the NBC Tower, the refrigeration system has three centrifugal chillers located in lower level 3 of the basement. Three cooling towers are on the roof of the building.

Chilled water cools from 58°F to 42°F (14. 4℃ to 5. 6℃) in the evaporator when the refrigerant is evaporated. The refrigerant is then compressed to the condensing pressure in the centrifugal compressor and is condensed in liquid form in the condenser, ready for evaporation in the evaporator.

There is no boiler in the central plant of the NBC Tower. To compensate heat loss in the perimeter zone, heat energy is provided by the warm plenum air and the electric heating coils in the fan-powered boxes.

Control System

Modern air conditioning control systems for the air and water systems and for the central plant consist of electronic sensors, microprocessor-operated and -controlled modules that can analyze and perform calculations from both digital and analog input signals, i. e. , in the form of a continuous variable. Control systems using digital signals compatible with the microprocessor are called direct digital control (DDC) systems (Unit 15). Outputs from the control modules often actuate dampers, valves, and relays by means of pneumatic actuators in large buildings and by means of electric actuators for small projects.

In the NBC Tower, the HVACR system is monitored and controlled by a microprocessor-based DDC system. The DDC controllers regulate the air-handling units and the terminals. Both communicate with the central operating station through interface modules. In case of emergency, the fire protection system detects alarm conditions. The central operating station gives emergency directions to the occupants, operates the HVACR system in a smoke control mode, and actuates the sprinkler water system.

Air, Water, Refrigeration, and Heating Systems

Air, water, refrigeration, heating, and control systems are the subsystems of an air conditioning or HVACR system. Air systems are often called secondary systems. Heating and refrigeration systems are sometimes called primary systems.

Central hydronic and space conditioning air conditioning systems both have air, water, refrigeration, heating, and control systems. The water system in a space conditioning system may be a chilled /hot water system. It also could be a centralized water system to absorb heat from the condenser during cooling, or provide heat for the evaporator during heating.

For a unitary packaged system, it consists of mainly air, refrigeration, and control systems. The heating system is usually one of the components in the air system. Sometimes a separate baseboard hot water heating system is employed in the perimeter zone.

An evaporative-cooling system always has an air system, a water system, and a control system. A separate heating system is often employed for winter heating.

In an individual room air conditioning system, air and refrigeration systems are installed in indoor and outdoor compartments with their own control systems. The heating system is often a component of the supply air chamber in the room air conditioner. It can be also a centralized hot water heating system in a PTAC system.

Air conditioning or HVACR systems are therefore often first described and analyzed through their subsystems and main components, such as air, water, heating, cooling/ refrigeration, and control systems. Air conditioning system classification, system operating characteristics, and system selection must take into account the whole system.

Among the air, water, and refrigeration systems, the air system conditions the air, controls and maintains the required indoor environment, and has direct contact with the occupants and the manufacturing processes. These are the reasons why the operating characteristics of an air conditioning system are essentially represented by its air system.

7.3 Troubleshooting Air-conditioning Systems

Troubleshooting an air-conditioning system correctly will result in the proper repair for the unit in a timely fashion. It is discouraging for the customer and the technician if multiple return visits are required to correct a malfunctioning unit. The trouble described by the customer may initially help you to determine which course of action needs to be taken. It will save you time to take a few minutes to discuss the symptoms with them. They may be able to provide you with valuable information that may lead you to the problem. Most manufacturers supply troubleshooting charts for their equipment. After speaking with the customer, a good next step is to review these to assist in troubleshooting the system. A quick simple fix may not always be the correct answer. For example, the unit may be tripped out due to a stuck high-water-level switch on the evaporator condensate drip pan.

Simply resetting the unit may bring it back on line, but if it trips out again you will be called to come back to the residence a second time. Before leaving the job, the drain pipe should be checked for an obstruction.

To avoid these types of situations, it is best to take the proper steps in evaluating the malfunction. Collect information about the problem. Read and calculate the system's vital signs, such as suction and discharge pressures. Compare the measured values to the expected manufacturer's recommendations. Consult the manufacturer's troubleshooting aids, if available, along with any company-specific troubleshooting guides.

Electrical testing is generally a first step in any troubleshooting sequence. Make sure that you properly identify the power supply to the unit. Do not assume that the power is always shut off at the breaker. Sometimes breakers are mislabeled. Always use your meter to check for ground faults before energizing the unit to be serviced. At all times, use your meter to check for voltage before physically disconnecting wires or opening control boxes.

System Familiarization

The air-conditioning system design may affect the way you troubleshoot the system. It is important to familiarize yourself with the system components and their location. Is the evaporator coil located in an indoor air-handling unit or is it installed in conjunction with a furnace? Is the air handler a verticalupflow, vertical downflow, or horizontal unit, and where is it located? Where are the system controls and switches located? What type of safety controls does the unit have? Check the capacity ratings for the system. A split system has a fan coil unit with a refrigeration coil located on the inside of the building and an outside-mounted air-cooled condensing unit, connected together with refrigerant piping.

Basically, there are three types of problems: trouble with the air system, the electrical system, or mechanical components. Within these there is much overlap, so whatever the nature of the problem, it is good practice to follow a logical, structured, systematic approach. In this manner, the correct solution is usually found in the shortest possible time.

Air System Problems

The primary problem that can occur in an air system is the reduction in airflow. Air-handling systems do not suddenly increase in capacity—that is, increase the amount of air across the coil. On the other hand, the refrigeration system does not suddenly increase in heat-transfer ability. First remove the panel to access the direct expansion (DX) coil in the air handler so that you can inspect the coil for dirt and blockage. On many systems, removing a panel would change the airflow across the coil and change the return and supply reading. Take the temperatures of the return air as it enters and then the supply air as it leaves the coil. The difference between these two temperatures is referred to as the temperature drop or temperature difference of the air across the DX coil.

A sling psychrometer is used to measure the return-air dry-bulb and wet-bulb temperatures needed to determine the relative humidity. Many electronic psychrometers are able to display relative humidity directly. A chart supplied by the DX coil manufacturer is then used to determine the expected temperature drop. As an example from the chart, return air at a condition of 68℉ and 30 percent relative humidity should have a 24℉ air temperature drop across the coil.

This measurement will help to determine if the problem is a result of improper airflow or a refrigeration system error. If the actual air temperature drop is greater than the required temperature drop, then the air quantity has been reduced. In this case, look for problems in the air-handling system. This could be due to dirty air filters, a dirty evaporator coil, a problem with the blower, or an unusual restriction in the duct system.

Air Filters

Because this is the most common problem of air failure, check the filtering system first. Air filters of the throwaway type should be replaced at least once a month.

Blower Motor and Drive

Check the blower motor and drive in the case of belt-driven blowers to make sure that both the blower motor and blower bearing are properly lubricated and operating freely. The blower drive belt must be in good condition and properly adjusted. Cracked or heavily glazed belts must be replaced. Heavy glazing can be caused by too much tension on the belt, driving the belt down into the pulleys. Proper adjustment requires the ability to depress the belt midway between the pulleys approximately 1 in for each 12 in between pulley centers.

The blower wheel should be clean. Dirt accumulation can sometimes fill in the area on a cupped blade, allowing it to spin freely but substantially reducing the airflow. If the wheel is dirty or has mold buildup, it must be removed and cleaned. Attempting to clean the wheel in place is never recommended. Do not try brushing only, because a poor cleaning job will cause an imbalance to occur on the wheel. Extreme vibration and noise will result. This could cause deterioration of the wheel, damage to the belt, and damage to the motor.

Unusual Restrictions in Duct Systems

Placing furniture or carpeting over return grilles reduces the air available for the blower to handle. Shutting off the air to unused areas will reduce the air over the coil. Covering a return-air grille to reduce the noise from the centrally located furnace or air handler may reduce the objectionable noise, but it also drastically affects the operation of the system by reducing air quantity. The condition of the grille may also indicate potential problems. Water stains can be caused by improper humidity conditions. Soot stains may be a result of an improperly functioning furnace's exhaust leaking back into the air-supply line. The collapse of the return-air duct system will affect the entire duct system performance. Air leaks in the return duct will raise the return-air temperature and reduce the temperature drop across the coil. Look for pinched ducts, sharp bends, and unnecessary duct length.

Air-distribution systems installed before the change in standards may have cloth or unapproved duct tape. These systems may need to be sealed properly for the central residential air-conditioning system to function properly. The technician should give particular attention to the integrity of the sealing of all joints and the insulation and tape if externally wrapped.

Measure the difference in temperature between the returnair at the grille compared to the return-air temperature as it enters the unit. This difference should not exceed 2°F. If it does, the return duct needs to be insulated or there may be leak openings in the duct that need to be sealed.

Some older fabric air-handler vibration dampers are made from asbestos. Always follow the proper procedures for handling asbestos materials. Never rip or tear the asbestos, and always wear an approved respirator.

Condensate Drain Pans

Condensate drain pans are located below the DX coil to collect any moisture that condenses from the air and drips off the coil. Depending on the humidity conditions, the amount of water drained away can be considerable. The condensate that collects will build up and overflow the pan if the drain becomes plugged or offers restricted flow. This is of particular concern with attic-mounted units because the condensate will spill over and damage the ceiling. Some condensate drip pans have condensate pumps that will stop and start automatically and are controlled by a float switch activated by the water level. Even if the condensate drain pan is empty, any signs of prior spillover should be investigated further.

Adjusting the Airflow

Most manufacturers supply operating data on their equipment (Table 7-1), indicating the total, sensible, and latent heat-removal rating at various outdoor dry-bulb and indoor wet-bulb temperatures at specific static pressures. For these same conditions, they supply the operating suction and discharge pressures of the equipment. This information is given so that the technician can match the actual conditions on the job with the performance conditions shown on the manufacturer's chart. Notice that in Table 7-1 a system with a low external static pressure of 0.30 would have an airflow of 625 CFM on LOW. But on a system with a higher external static pressure of 0.70, the blower would have to be set to HIGH to obtain the same airflow.

Comparison of water and air. **Table 7-1**

Parameter	Air	Water
Specific heat, Btu/ lb · °F	0.243	1.0
Density, at 68°F, lb/ ft^3	0.075	62.4
Heat capacity of fluid at 68°F, Btu/ft^3 · °F	0.018	62.4

Most manufacturers recommend that a system's airflow be set so there is approximately 400 CFM per ton of air conditioning. In areas that have high relative humidity, a slightly slower fan speed can be used to aid in the dehumidification. In arid areas, the fan speed can be increased to provide more sensible cooling. As a rule of thumb, the fan speed should not be adjusted more than 10 percent above or below the manufacturer's recommendations. Excessively slow fan speeds can allow the evaporator coil to freeze up under light loads, and excessively high fan speeds can put too large a load on the condenser under heavy load conditions.

Electrical Problems

Since the greatest numbers of malfunction problems are electrical, it is common practice to perform electrical troubleshooting (including controls) before mechanical troubleshooting. If the problem is mechanical, the electrical check will usually point the technician in that direction. If the system will not operate at all, it is probably an electrical problem that must be found and corrected.

Before beginning any troubleshooting, you must locate the manufacturer's troubleshooting guide for the equipment you are working on Manufacturers have developed troubleshooting techniques that when followed will result in accurate and rapid location of the problem. These charts may seem complex and difficult to follow if you look at the entire chart, so to properly use one of these charts, start at the beginning and follow each and every step. Do not simply jump ahead assuming that you know what the answer is. Do the test and report the results, and then move to the next test as indicated by the manufacturer's guide.

Electrical Operating Sequence

The operating sequence is usually supplied by the manufacturer in the service instructions, or it can be determined by the technician by studying the schematic wiring diagram. The functions of the operating and nonoperating equipment are determined by examination and testing. Necessary test instruments include the volt ohm meter, the clamp-on ammeter, the capacitor tester, and the temperature analyzer. The power circuit is the first to be examined, because power must be available to operate the loads. For example, on a refrigeration system with an air-cooled condenser, the three principal loads that must be energized are the compressor motor, the condenser fan motor, and the evaporator fan motor. Before proceeding with anything else, the technician must be certain that the proper voltage is being supplied to the loads. The supply-line voltage should be tested with a voltmeter and then compared to the manufacturer's recommended values.

If the voltage is correct, then the circuit will need to be tested further. Before energizing the circuit, always use your meter to check for ground faults. Once energized, the voltage for each load in the circuit can be tested. A clampon ammeter can be used to check the current draw for each load one at a time. The measured current should be compared to the current rating listed on the motor nameplate. As an example, if the current draw is too high, this may indicate a mechanical problem such as a stuck fan or seized compressor.

In troubleshooting, when a load is not working, the technician must determine whether the problem is in the load itself or in the switches that control the load. If the proper supply voltage is available but the compressor or fan motor does not run, then there may be a fault in the control circuit and the control circuit voltage should be tested. Generally the control circuit voltage will be much lower than the line voltage. Each load, compressor, or fan is generally controlled by a relay switch operated by an electromagnetic coil energized by the control circuit.

Use your meter to check for voltage before physically disconnecting wires or opening control boxes. Always test to determine if the fan or motor is tripped out on a safety switch before replacing components. When checking components with an ohmmeter, it is important that all power is disconnected and the component part is electrically isolated. Short circuits are usually due to faulty loads. If a faulty component is located, the job is still not complete. The technician should make a concerted effort to determine why or how the component failure occurred.

Installation and Service Instructions

The installation and service instructions supply a wide variety of information that the manufacturer believes is necessary to properly install and service the unit. This bulletin includes the wiring diagram, the sequence of operation, and any notes or cautions that need to be observed in using them.

Wiring Diagrams

Wiring diagrams usually consist of connection diagrams and schematic diagrams. The connection diagram shows the wires to the various electrical component terminals in their approximate location on the unit. This is the diagram that the technician must use to locate the test points. The schematic diagram separates each circuit to clearly indicate the function of switches that control each load. This is the diagram the technician uses to determine the sequence of operation for the system.

Wiring diagrams are often included inside the equipment panels. If the diagrams are not there, then contact the manufacturer. They are available from local suppliers and are often available from the manufacturers themselves through their Web sites. If you do not have a diagram, troubleshooting electrical circuits can be lengthy, time-consuming, and inaccurate.

Troubleshooting Tables

Troubleshooting tables are helpful as a guide to corrective action. By a process of elimination, such tables offer a quick way to solve a service problem. The process of elimination permits the technician to examine each suggested remedy and disregard ones that do not apply or are impractical; leaving only the solution(s) that fits the problem.

Fault Isolation Diagrams

A fault isolation diagram starts with a failure symptom and goes through a logical decision action process to isolate the failure.

Diagnostic Tests

Diagnostic tests can be conducted on electronic circuit boards, at points indicated by the manufacturer, to check voltages or other essential information critical to the operation of the unit.

Some electronically controlled systems have automatic testing features, which indicate by code number a malfunction in the operation of the equipment. Further tests are usually required to determine the action that is required.

Mechanical Problems

When the measured tempera ture drop across the DX coil is less than required, this means that the heat-removal capacity of the system has been reduced. This means that the amount of heat picked up in the coil plus the amount of motor heat added and the total rejected from the condenser is not the total heat quantity the unit is designed to handle. The problems associated with a system that starts and runs but does not produce satisfactory cooling can be simply divided into two categories: refrigerant quantity and refrigerant flow rate. To determine the problem, all the information listed in Table 7-2 must be measured. These results compared to normal operating results will generally identify the problem. The use of the word normal does not imply a fixed set of pressures and temperatures. These will vary with each make and model of the system. A few temperatures are fairly consistent throughout the industry and can be used for comparison. These are the DX coil operating tempera ture, the condensing unit condensing temperature, and the refrigerant subcooling.

These vital signs must also be modified according to the seasonal energy efficiency ratio (SEER)of the unit. The reason for this is that the amount of evaporation and condensing surface designed into the unit is directly related to the efficiency rating. A larger condensing surface results in a lower condensing temperature and a higher SEER. A larger evaporating surface results in a higher suction pressure and a higher SEER. The energy efficiency ratio is calculated by dividing the net capacity of the unit in BTU/hr by the watts input. Every central split cooling system manufactured in the United States today must have a SEER of at least 13.

Refrigerant Subcooling

The amount of subcooling produced in the condenser is affected by the quantity of refrigerant in the system. The temperature of the air entering the condenser and the load on the DX coil also has an effect on the amount of subcooling produced. Typically, it is desirable to have liquid subcooling of approximately 15°F～20°F.

If a refrigerant liquid line rises over 30 ft vertically, additional subcooling may be required to ensure that the metering device is receiving 100 percent liquid. The additional subcooling is required because of the pressure difference between the refrigerant at the condenser and pressure of the refrigerant at the metering device. This pressure drop is the result of the static head caused by the vertical lift of the refrigerant. Refer to the manufacturer's technical specifications to see what additional subcooling is required for unusually high vertical lifts.

Insufficient or Unbalance Load

Insufficient air over the DX coil would be indicated by a greater than desired temperature drop through the coil. An unbalanced load on the DX coil would also give the opposite indication—some of the circuits of the DX coil would be overloaded, while others would be lightly loaded. This would result in a mixture of air off the coil that would cause a reduced temperature drop of the air mixture. The lightly loaded sections of the DX coil would allow liquid refrigerant to leave the coil and enter the suction manifold and suction line.

The final step of installation that takes place at the residence is outside the control of the manufacturer. Many manufacturers have found that significant problems can exist in system installation, including the mismatching of equipment sizes. If you suspect that your refrigerant circuit problems are the result of mismatched equipment, call the equipment supplier or manufacturer and provide them the system component model numbers so they can inform you as to whether the system is mismatched in size and make recommendations to resolve the problems if a mismatched system exists.

In TEV systems, the liquid refrigerant passing the sensing bulb of the TEV would cause the valve to close down. This would reduce the operating temperature and capacity of the DX coil and lower the suction pressure. This reduction would be very pronounced. The DX coil operating superheat would be very low, probably zero, because of the liquid leaving some of the sections of the DX coil.

Discharge pressure (high side) would be low due to the reduced load on the compressor, reduced amount of refrigerant vapor pumped, and reduced heat load on the condenser. Condenser liquid subcooling would be higher than normal because of the reduction in refrigerant demand by the TEV. The condensing unit amperage draw would be down due to the reduced load.

In systems using fixed metering devices, the unbalanced load would produce a lower temperature drop of the air through the DX coil because the amount of refrigerant supplied by the fixed metering device would not be reduced; therefore, the system pressure (boiling point) would be approximately the same.

The DX coil superheat would drop to zero with liquid refrigerant flooding into the suction line. Under extreme cases of imbalance, liquid returning to the compressor could cause compressor damage. The reduction in heat gathered in the DX coil and the decrease of refrigerant vapor to the compressor will lower the load on the compressor. The compressor discharge pressure will be reduced.

The flow rate of the refrigerant will only be slightly reduced because of the lower discharge pressure. The subcooling of the refrigerant will be in the normal range. The amperage draw of the condensing unit will be slightly lower because of the reduced load on the compressor and reduction in head pressure.

Excessive Load

In the case of excessive load, the opposite effect exists. The temperature drop of the air through the coil will be less, because the unit cannot cool the air as much as it should. Air is moving through the coil at too high a velocity. There is also the possibility that the temperature of the air entering the coil is higher than the return air from the conditioned area. This could be from air leaks in the return duct system drawing hot air from unconditioned areas.

The excessive load raises the suction pressure. The refrigerant is evaporating at a rate faster than the pumping rate of the compressor. If the system uses a TEV, the superheat will be normal to slightly high. The valve will operate at a higher flow rate to attempt to maintain superheat settings. If the system uses fixed metering devices, the superheat will be high. The fixed metering devices cannot feed enough refrigerant to keep the DX coil fully active.

The discharge pressure will be high. The compressor will pump more vapor because of the increase in suction pressure. The condenser must handle more heat and will develop a higher condensing temperature. A higher condensing temperature means a greater discharge pressure.

The quantity of liquid in the system has not changed, nor is the refrigerant flow restricted. The liquid subcooling will be in the normal range. The amperage draw of the unit will be high because of the additional load on the compressor.

Low Ambient Temperature

In this case, the condenser heat transfer rate is excessive, producing an excessively low discharge pressure. As a result, the suction pressure will be low because the amount of refrigerant through the metering device will be reduced. This reduction will reduce the amount of liquid refrigerant supplied to the DX coil. The coil will produce less vapor and the suction pressure drops.

Often air-conditioning and refrigerant equipment dies at night. This is frequently because of the lower ambient temperatures experienced in the evening. Low ambient operation of refrigeration and air-conditioning compressors can result in a liquid floodback or slugging of the compressor. If a system is to be operated on a regular basis during low ambient conditions, it should be equipped with a low ambient kit to protect the compressor.

The decrease in the refrigerant flow rate into the coil reduces the amount of active coil, and a higher superheat results. In addition, the reduced system capacity will decrease the amount of heat removed from the air. There will be a higher temperature and relative humidity in the conditioned area, and the discharge pressure will be low. This starts a reduction in system capacity. The amount of subcooling of the liquid will be in the normal range. The quantity of liquid in the condenser will be higher, but the heat transfer rate of the evaporator is less. The amperage draw of the condensing unit will be less because the compressor is doing less work.

The amount of drop in the condenser ambient air temperature that the air-conditioning system will tolerate depends on the type of pressure-reducing device in the system. Systems using fixed metering devices will have a gradual reduction in capacity as the outside ambient drops from 95°F. This gradual reduction occurs down to 65°F. Systemsthat use a TEV will maintain higher capacity down to an ambient temperature of 47°F. Below these temperatures the capacity loss is drastic, and some means of maintaining discharge pressure must be employed to prevent the evaporator temperature from dropping below freezing. The most reliable means is control of air through the condenser via dampers in the airstream, the condenser fan cycling on and off, a variable-speed condenser fan, or some combination of these components.

High Ambient Temperature

When the outside air temperature rises on a hot day, the temperature of the air entering the condenser will be higher. This also increases the condensing temperature and pressure of the refrigerant vapor. The suction pressure will also be high because the pumping efficiency of the compressor is reduced. There will also be less liquid-line subcooling, which will increase the amount of flash gas across the metering device, further reducing the system efficiency. Due to the high ambient temperature, the discharge pressure will be high. There will be less liquid refrigerant in the condenser and reduced liquid subcooling. The system will run less efficiently and therefore will require more power, so the amperage draw of the condensing unit will be high.

The amount of superheat produced in the coil will be different in a TEV system as compared to a system using a fixed metering device. In the TEV system, the valve will maintain superheat close to the limits of its adjustment range even though the actual temperatures involved will be higher.

In a fixed metering device system, the amount of superheat produced in the coil is the reverse of the temperature of the air through the condenser. The flow rate through the fixed metering device is directly affected by discharge pressure. The higher air temperature will result in a higher discharge pressure and a higher flow rate. As a result of the higher flow rate, the amount of subcooling in the condenser is lower.

Table 7-2 shows the superheat that will be developed in a properly charged air-conditioning system using a fixed metering device. Do not attempt to charge a fixed metering device system below 65°F, as system operating characteristics become very erratic.

The effects of outdoor (ambient) temperature on superheat. **Table 7-2**

Outdoor Air Temperature Entering Condenser Coil (°F)	Superheat (°F)
65	30
75	25
80	20
85	18
90	15
95	10
105 and above	5

Refrigerant Undercharge

With a shortage of refrigerant in the system, less liquid refrigerant enters the evaporator coil to pick up heat, creating a lower suction pressure. The smaller quantity of liquid supplied to the coil results in less active surface for the evaporator coil for vaporizing the liquid refrigerant and more surface to raise vapor temperature. The superheat will be high. There will beless vapor for the compressor to handle and less heat for the condenser to reject, leading to a lower condensing temperature and discharge pressure.

The compressor in an air-conditioning system is cooled primarily by the cool returning suction gas. Compressors that are low on charge can have a much higher operating temperature. The temperature can be high enough so that the motor windings begin to break down. As this occurs, the motor can ultimately short out, resulting in a compressor change-out. If an air-conditioning system has a leak, it must be located so that a low refrigerant charge can be avoided.

The amount of subcooling will be below normal to zero, depending on the amount of undercharge. The system operation is usually not affected very seriously until the subcooling is zero and hot gas starts to leave the condenser together with the liquid refrigerant. The amperage draw of the condensing unit will be slightly less than normal.

Unit 8 Heating Systems

For thousands of years man has used fire for warmth. In the beginning interior heating was just an open fire in a cave with a hole at the top. Later, fires were contained in hearths or sunken beneath the floor. Eventually, chimneys were added which made for better heating, comfort, health, and safety and also allowed individuals to have private rooms. Next, came stoves usually made of brick, earthenware, or tile. In the 1700s, Benjamin Franklin improved the stove, the first steam heating system was developed, and a furnace for warm-air heating used a system of pipes and flues and heated the spaces by gravity flow. In the 1800s, high-speed centrifugal fans and axial flow fans with small, alternating current electric motors became available and high-pressure steam heating systems were first used. The 1900s brought the Scotch marine boiler and positive-pressure hydronic circulating pumps that forced hot water through the heating system. The heating terminals were hot water radiators, which were long, low, and narrow, and allowed for inconspicuous heating when compared to tall, bulky, out-in-the-open steam radiators.

Centrifugal fans were added to furnaces in the 1900s to make forced-air heating systems similar to the ones used in today's residential and commercial systems. Larger commercial heating systems most often used today are low temperature (with boiler water temperatures generally in the range of 170°F to 200°F) or low pressure steam heating systems using boilers which operate around 15 pounds pressure [pounds per square inch, gauge (or psig) which is equivalent to approximately 30 pounds per square inch, absolute pressure, (psia)]; this is a 250°F steam temperature.

Combustion is defined as a chemical reaction between a fossil fuel such as coal, natural gas, liquid petroleum gas, or fuel oil, and oxygen. Fossil fuels consist mainly of hydrogen and carbon molecules. These fuels also contain minute quantities of other substances (such as sulfur) which are considered impurities. When combustion takes place, the hydrogen and the carbon in the fuel combine with the oxygen in the air to form water vapor and carbon dioxide.

If the conditions are ideal, the fuel-to-air ratio is controlled at an optimum level, and the heat energy released is captured and used to the greatest practical extent. Complete combustion (a condition in which all the carbon and hydrogen in the fuel would be combined with all the oxygen in the air) is a theoretical concept and cannot be attained in HVACR equipment. Therefore, what is attainable is called incomplete combustion. The products of incomplete combustion may include unburned carbon in the form of smoke and soot, carbon monoxide (a poisonous gas), as well as carbon dioxide and water.

8.1 Boiler Heating Systems

Boiler heating systems provide heat to designated areas by transporting heat energy generated in the boiler. The two types of boiler heating systems are water heating and steam heating. The difference in heating systems is the medium used to transport the energy from the boiler to the area to be heated. Water is used to transport heat energy in the water heating system and steam is used to transport energy in the steam heating system.

Water Heating System

Water heating systems transport heat by circulating heated water to a designated area. Heat is released from the water as it flows through a coil (heat exchanger). After heat is released, the water returns to the boiler to be reheated and recirculated. Low temperature water boiler are ≤250°F and high temperature water boilers are greater than>250°F water temperature.

In the water heating system in Figure 8-1 the return water (HWR) from the heating coil enters the boiler at 180°F and is heated to 200°F. The 200°F supply water (HWS) is pumped from the boiler by the hot water pump (HWP) and enters the heating coil. The water flows through the coil giving off heat. Mixed air passes over the coil tubes and fins picking up heat from the water and is drawn through the supply air fan to maintain room air temperature at 70°F. The air temperature leaving the coil is 105°F. The return air from the conditioned space mixes with the outside air at 30°F to get a mixed air temperature of 62°F. You can do this calculation by determining the amount of supply air to the condi-

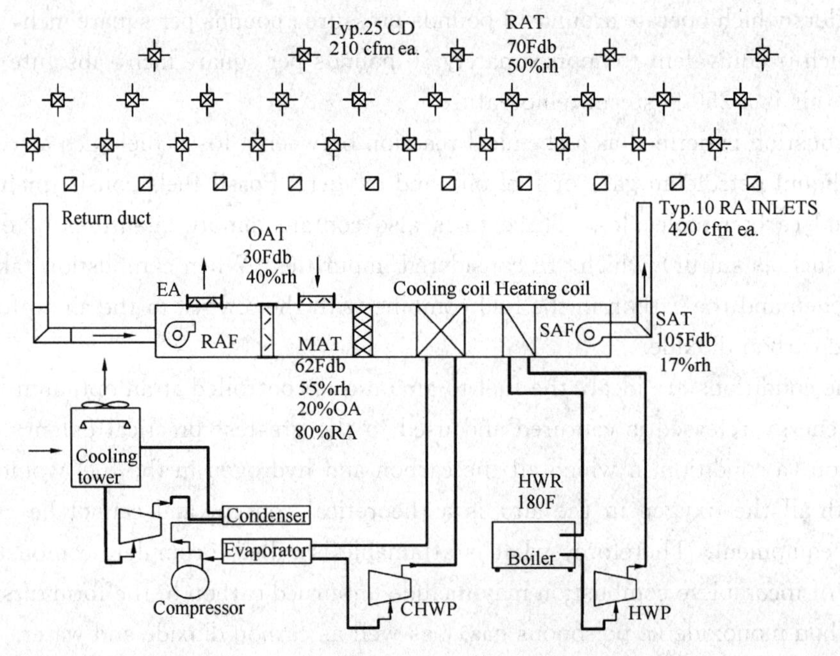

Figure 8-1 Water heating system and air distribution.

tioned space and the amount of return air. Calculate the percent return air and subtract from 100% to get percent of outside air. Use the MAT equation at the end of this chapter to find mixed air temperature. The 62°F mixed air temperature enters the heating coil and is heated to 105°F supply air temperature. Additional notes for Figure 8-1: The water pump should be installed on the leaving water (aka HWS) side of the boiler so it pumps the water out of the boiler. "Mixed air" may be called "conditioned air" after it leaves any energized (operating) heating or cooling coil. A "draw-through" system is when the fan is after the coil(s), if the fan is installed before the coil(s) it is a "blow-through" system.

When water is heated in the boiler some air is entrained in the water and is piped along with the water to the heat exchangers (coils). This air is removed from the water through automatic or manual air vents and air separators in the system as shown in Figure 8-2.

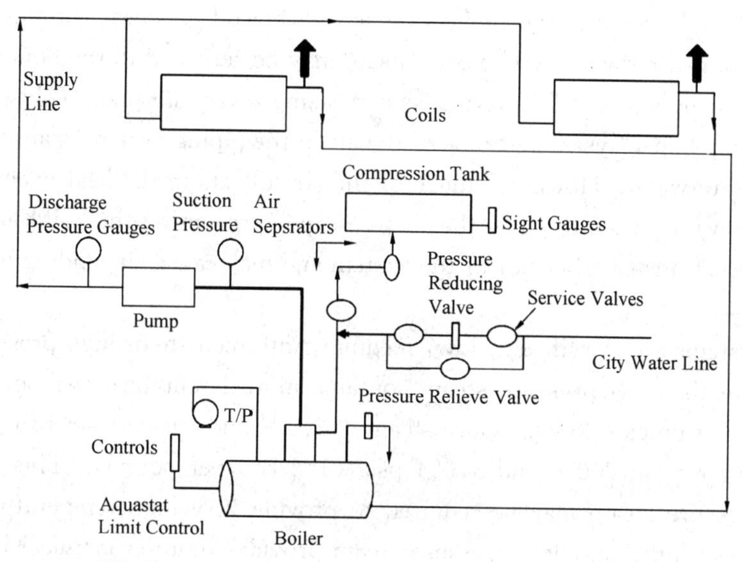

Figure 8-2 Water heating system.

The boiling point (aka boiling temperature) of water can be changed by changing the pressure on the water. If the pressure is to be changed the water must be in a closed vessel (boiler) and then the water can be boiled at a temperature of 212°F or 250°F or any other temperature desired. If the pressure on the water in the boiler is 14.7 psia the boiling temperature is 212°F.

Sea level barometric pressure is 14.7 pounds per inch absolute (psia). Sea level barometric pressure is also 0 pounds per square inch gauge (psig). To change from pressure absolute to pressure gauge add or subtract 14.7. The equation is

$$\text{psia} = \text{psig} + 14.7$$

Sea level barometric pressure is also stated as 29.92 inches of mercury (inches Hg or in. Hg). The equations for changing inches of mercury to psi are:

$$1'' \text{ Hg} = 0.49 \text{ psi}$$

$$1 \text{ psi} = 2.04''$$

For estimation purposes only, round off 1" Hg to 0.5 psi or round off 1 psi to 2" Hg. Sometimes sea level barometric pressure, for estimation purposes only, is rounded off to 15 psia and 30" Hg. Also, it's common for the terms "pressure absolute" and "pressure gauge" to be changed around and said or written as "absolute pressure" and "gauge pressure."

Steam Heating System

In a steam heating system, water enters a heat conversion unit (the boiler, a closed vessel) and is changed into steam. When the water is boiled, some air in the water is released into the steam and is piped along with the steam to the heat exchanger (coil). At the coil, heat is released into the air flowing across the coil. As heat is removed the steam changes into condensate water. The condensate may be returned to the boiler by a gravity return system or by a mechanical return system using a vacuum pump (closed system) or condensate pump (open system). Some of the air in the piping system is absorbed back into the condensate water. However, much of the air collects in the heat exchanger. Steam traps (see below) are used to allow the air to escape, preventing the build-up of air which reduces the heat transfer efficiency of the system and may cause air binding in the heat exchanger.

Steam systems are classified as low, medium, intermediate or high pressure. It is important to note that low pressure steam contains more latent heat per pound than high pressure steam. Compare low pressure steam at 250°F and 30psia (946 Btu per pound) to high pressure steam at 700°F and 3,094 psia (172 Btu per pound). This indicates that while high pressure steam may be required to provide very high temperatures and pressures for process functions, low pressure steam provides more economical heating operation.

Steam has some design and operating advantages over water for large heating systems. Heat capacity is one, prime mover is another. Steam at 212°F when condensed releases or gives up approximately 1000 Btu per pound. On the other hand, a water heating system with supply water temperatures at 200°F and return water temperatures at 180°F only releases 20 Btu per pound (1 Btu/lb • °F). A water system uses a motor and pump to overcome system resistance (in pipes, valves, heat exchangers, etc.) to circulate the water through the system. Steam, based on its operating pressure, flows throughout the system on its own. Therefore, a motorized circulating pump is not needed although some systems do use small condensate return pumps.

Heat loss occurs through pipe radiation losses in both steam and water heating systems and at steam traps in steam heating systems. There are energy conservation opportunities in maintaining traps and proper insulation on steam and eater pipes.

Boiling Temperatures and Pressures

In an open vessel, at standard atmospheric pressure (sea level), water vaporizes or boils into steam at a temperature of 212°F. But the boiling temperature of water, or any liquid, is not constant. The boiling temperature can be changed by changing the pressure on the liquid. If the pressure is to be changed, the liquid must be in a closed vessel. In a water or steam heating system the vessel is the boiler. When the water is in the boiler it can be boiled at a temperature of 100°F or 250°F or 300°F as easily as at 212°F. The only requirement is that the pressure in the boiler be changed to the one corresponding to the desired boiling point. For instance, if the pressure in the boiler is 0.95 pounds per square inch absolute (psia), the boiling temperature of the water will be 100°F. If the pressure is raised to 14.7 psia, the boiling temperature is raised to 212°F. If the pressure is raised again to 67 psia, the temperature is correspondingly raised to 300°F. A common low pressure HVACR steam heating system will operate at 15 pounds per square inch gauge pressure (psig), which is a pressure of 30 psia and a temperature of 250°F.

The amount of heat required to bring the water to its boiling temperature is its sensible heat. Additional heat is then required for the change of state from water to seam. This addition of heat is steam's latent heat content or "latent heat of vaporization." To vaporize one pound of water at 212F requires 970 Btu. The amount of heat required to bring water from any temperature to steam is called "total heat." It is the sum of the sensible heat and latent heat. The total heat required to convert one pound of water at 32°F to one pound of steam at 212°F is 1,150F Btu. The calculation is as follows: the heat required to raise one pound of water at 32°F to water at 212°F is 180 Btu of sensible heat. 970 Btu of latent heat is added to one pound of water at 212°F to convert it to one pound of 212°F steam. Notice that the latent heat is over 5 times greater than sensible heat.

$$\text{The latent heat} = 180\ \text{Btu} \times 5.39 = 970\ \text{Btu}$$

The total heat is:

$$\text{The total heat} = \text{The sensible heat} + \text{The latent heat} = 180 + 970 = 1150\ \text{Btu}$$

See Figure 8-3.

Points in Figure 8-3 correspond to the following:

Point 1-One pound of ice(a solid)at 0°F

Point 1 to Point 2—16 Btu of sensible heat added to raise the temperature of the ice from 0°F to 32°F. Specific heat of ice is 0.5 Btu/lb • °F. Diagonal line from point to point is sensible heat and horizontal line from point to point is latent heat.

Point 2 to Point 3—Ice changing to water (a liquid) at 32°F. It takes 144 Btu of latent heat to change one pound of ice to one pound of water. Total heat added is 160 Btu from point 1 to point 3.

Point 3 to Point 4—180 Btu of sensible heat added to raise the temperature of the water from 32°F to 212°F. Specific heat of water is 1.0 Btu/lb • °F. Total heat added is 340 Btu from point 1 to point 4.

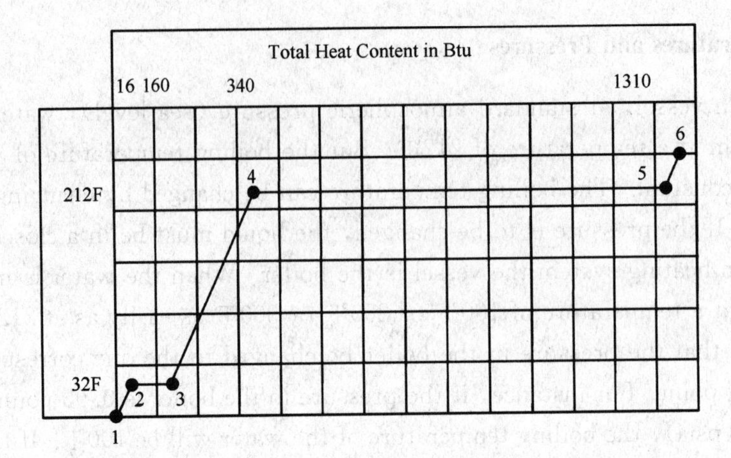

Figure 8-3 Btu change from one pound ice to water to steam to superheated steam.

Point 4 to Point 5—Water changing to steam (a vapor) at 212°F. It takes 970 Btu of latent heat to change one pound of water to one pound of steam. Total heat added is 1,310 Btu from point 1 to point 5.

Point 5 to Point 6—X amount of Btu of sensible heat added to raise the temperature of the steam from 212°F toX°F. This is called superheating the steam and the result is "superheated steam." For example, if the final temperature of the superheated steam is 250°F then 19 Btu of sensible heat would have to be added.

$$250°F - 212°F = 38°F$$

$$38°F \times 0.5\ Btu/lb \cdot °F \text{ specific heat for steam} \times 1\ lb \text{ of steam} = 19\ Btu$$

8.2 Heating Components

Boilers

Boilers are used in both hot water heating systems and steam heating systems. The hot water heating systems most often encountered in HVACR work will be low temperature systems with boiler water temperatures generally in the range of 170°F-200°F Most of the steam heating systems will use low pressure steam, operating at 15 psig (30psia, and 250°F). There are a great many types and classifications of boilers. Boilers can be classified by size, construction, appearance, original usage, and fuel used. Fossil-fuel boilers will be either natural gas-fired, liquid petroleum gas-fired or oil-fired. Some boilers are set up so that the operating fuel can be switched to natural gas (NG), liquid petroleum gas (LPG) or oil, depending on the fuel price and availability. The construction of boiler remains basically the same whether they're water boilers or steam boilers. However, water or steam boilers are divided by their internal construction into fire tube or water tube boilers.

Fire Tube Boiler

A fire tube boiler, as the name suggests, has the hot flue gases from the combustion chamber-the chamber in which combustion takes place-passing through tubes and out the boiler stack. These tubes are surrounded by water. The heat from the hot gases transfers through the walls of the tubes and heats the water. Fire tube boilers may be further classified as externally fired, meaning that the fire is entirely external to the boiler or they may be classified as internally fired, in which case, the fire is enclosed entirely within the steel shell of the boiler. Two other classifications of fire tube boilers are wet-back or dry-back. This refers to the compartment at the end of the combustion chamber. This compartment is used as an insulating plenum so that the heat from the combustion chamber, which can be several thousand degrees, does not reach the boiler's steel jacket. If the compartment is filled with water it is known as a wet-back boiler and conversely, if the compartment contains only air is called a dry-back boiler.

Still another grouping of fire tube boilers is by appearance or usage. The two common types used today in HVACR heating systems are the marine or Scotch marine boiler and the firebox boiler. The marine boiler was originally used on steam ships and is long and cylindrical in shape. The fire box boiler has a rectangular shape, almost to the point of being square. A Scotch marine fire tube boiler has the flame in the furnace and the combustion gases inside the tubes. The furnace and tubes are within a larger vessel, which contains the water and steam. Fire tube boilers are also identified by the number of passes that the flue gases take through the tubes. Boilers are classified as two-, three- or four-pass. The combustion chamber is considered the first pass. Therefore, a two-pass boiler would have one-pass down the combustion chamber looping around and the second pass coming back to the front of the boiler and out the stack. A three-pass boiler would have an additional row of tubes for the gas to pass through going to the back of the boiler and out the stack. A four-pass boiler would have yet another additional row of tubes for the gas to pass through going to the front of the boiler and out the stack. An easy way to recognize a two-, three- or four-pass boiler is by the location of the stack. A two- or four-pass boiler will have the stack at the front, while a three-pass boiler will have the stack at the back.

Fire tube boilers are available for low and high pressure steam, or hot water applications. The size range is from 15 to 1,500 boiler horsepower (a boiler horsepower is 33,475Btuh). HVACR fire tube boilers are typically used for low pressure applications.

Water Tube Boiler

In a water tube boiler, the water is in the tubes while the fire is under the tubes. The hot flue gases pass around and between the tubes, heating the water and then out the boiler stack. Most of the water tube boilers used in heating systems today are rectangular in shape with the stack coming off the top, in the middle of the shell. Water tube boilers produce steam or hot water for industrial processes, commercial applications or other modest-size applications. They are used less frequently for comfort heating applications. Water tube boilers typically range from 25 boiler horsepower (836,875 Btuh or 836.88 MBh) to

250 boiler horsepower (8,368,750 Btuh, or 8368.75 MBh or 8.37 MMBh).

Electric Boiler

Electric boilers produce heat by electricity and operate at up to 16,000 volts. Electric boilers are typically compact, clean and quiet. They have replaceable heating elements, either electrode or resistance-coil. With the electrode type boiler, the heat is generated by electric current flowing from one electrode to another electrode through the boiler water. Resistance-coil electric boilers have the electricity flowing through a coiled conductor similar to an electric space heater. Resistance created by the coiler conductor generates heat. Resistance-coil electric boilers are not as common as electrode electric boilers.

Electric boilers are an alternative to oil or gas boilers where these boilers are restricted by emission regulation and in areas where the cost of electric power is minimal. Electric boiler can be fire tube or water tube and supply low to high pressure steam or hot water. Sizes range from 9 kW to 3,375 kW output, which is 30,708Btuh to 11,515,500 Btuh (1 kW = 3,412 Btuh).

For a better understanding of boiler construction and operation, let's examine a four-pass, internally fired, fire tube, natural gas-fired, forced-draft, marine, wet-back boiler. The boiler consists of a cylindrical steel shell which is called the pressure vessel. It is covered with several inches of insulation to reduce heat loss. The insulation is then covered with an outer metal jacket to prevent damage to the insulation. Some of the other components are a burner, a forced-draft fan and various controls.

When the boiler is started it will go through a purge cycle in which the draft fan at the front of the boiler will force air through the combustion chamber and out the stack at the front of the boiler. This purges unwanted combustibles that might be in combustion chamber. An electrical signal from a control circuit will open the pilot valve allowing natural gas to flow to the burner pilot light. A fame detector will verify that the pilot is lit and gas will then be supplied to the main burner. The draft fan forces air into the combustion chamber and combustion takes place. The hot combustion gases flow down the chamber and into the tubes for the second pass back to the front of the boiler. As the gases pass through the tubes they are giving up heat into the water. The gases enter into the front chamber of the boiler, called the header, and make another loop to the back of the boiler for the third pass. The fourth pass brings the hot gases back to the front of the boiler and out the stack. The temperature in the combustion chamber is several thousand degrees while the temperature of the gases exiting the stack should be about 320 degrees (or 150 degrees above the medium temperature).

A high fuel-to-air ratio causes soothing and lowers boiler efficiency. In certain conditions, it may also be dangerous if there's not enough air for "complete combustion" and dilution of the fuel. An improperly adjusted burner, a blocked exhaust stack, the blower or damper set incorrectly, or any condition which results in a negative pressure in the boiler room, can cause a high fuel-to-air condition. A negative pressure in the boiler room can be the result of one or a combination of conditions such as an exhaust fan creating a negative pressure in the boiler room, a restricted combustion air louver into the boiler room, or even adverse wind conditions.

High air to fuel ratios also reduce boiler efficiency. If too much air is brought in (excess air), the hot gases are diluted too much and move too fast through the tubes before proper heat transfer can occur. High air volumes are typically caused by improper blower or damper settings.

Steam Traps

Steam traps are installed in locations where condensate is formed and collects, for example, all low points, below heat exchangers and coils, at risers and expansion loops, at intervals along horizontal pipe runs, ahead of valves, at ends of mains, before pumps, etc. The purpose of a steam trap is to separate the steam (vapor) side of the heating system from the condensate (water) side. A steam trap collects condensate and allows the trapped condensate to be drained from the system, while still limiting the escape of steam. The condensate may be returned to the boiler by a gravity return system, a mechanical return system using a vacuum pump (closed system), or condensate pump (open system).

Condensate must be trapped and then drained immediately from the system. If it isn't, the operating efficiency of the system is reduced because the heat transfer rate is slowed. In addition, the buildup of condensate can cause physical damage to the system from"water hammer." Water hammer can occur in a steam distribution system when the condensate is allowed to accumulate on the bottom of horizontal pipes and is pushed along by the velocity of the steam passing over it. As the velocity increases, the condensate can form into a non-compressible slug of water. If this slug of water is suddenly stopped by a pipe fitting, bend, or valve the result is a shock wave which can, and often does, cause damage to the system (such as blowing strainers and valves apart).

Steam traps also allow air to escape. This prevents the build-up of air in the system which reduces the heat transfer efficiency of the system and may cause air binding in the heat exchanger. In a steam heating system, water enters a heat conversion unit such as a heat exchanger or a boiler and is changed into steam. When the water is boiled, some air in the water is also released into the steam and is moved along with the steam to the heating coils or other heat exchanger. As the heat is released at the heat exchangers (and through pipe radiation losses) the steam is changed into condensate water. Some of the air in the piping system is absorbed back into the water. However, much of the air collects in the heat exchanger and must be vented.

Steam traps are classified as thermostatic, mechanical or thermodynamic. Thermostatic traps sense the temperature difference between the steam and the condensate using an expanding bellows, bimetal strip or other sensor to operate a valve mechanism. Mechanical traps use a float to determine the condensate level in the trap and then operate a discharge valve to release the accumulated condensate. Some thermodynamic traps use a disc which closes to the high velocity steam and opens to the low velocity condensate. Other types will use an orifice which flashes the hot condensate into steam as the condensate passed through the orifice.

Burner

The function of the burner is to deliver, ignite and burn the proper mixture of air and fuel. The types of burners are varied and selection depends on the design of the boiler. Oil burners (Figure 4-8), except for small domestic types, deliver the fuel to the burner under pressure provided by the oil pump. The heavier oils, numbers 4, 5, and 6, generally require preheaing to lower their viscosity so that they can be pumped to the burner. In addition, all oils must be converted to a vapor before they can be burned. Large commercial and industrial burners use two steps to prepare the oil for burning. The first step is called atomization which is the reduction of the oil into very small droplets. The second step is vaporization which is accomplished by heating the droplets. Oil burners are classified by how they prepare the oil for burning such as vaporizing, atomizing or rotary. Oil burners use the same methods of delivering air to the combustion chamber as do gas burners. They are either natural-, forced- or induced-draft. Regardless of what type of burner is used, proper combustion depends on the correct ratio of fuel-to-air.

Gas burners are classified as atmospheric or mechanical-draft burners. Atmospheric burners are sub-classified as natural-draft or Venturi burners. Mechanical-draft burners are either forced- or induced-draft burners. A typical gas burner used on large industrial and commercial boilers is a burner with a fan or blower at the inlet. This type of burner, which is a force-draft burner, is called a power burner (Figure 4-10). It uses the blower to provide combustion air to the burner and the combustion chamber under pressure and in the proper mixture with the gas over the full range of firing from minimum to maximum. Another type of gas burner uses a blower at the outlet of the combustion chamber to create a slight partial vacuum within the chamber. This causes a suction which draws air into the chamber. This type of burner is an induced-draft burner.

8.3 Calculations for Heating System

Looking at the heating system, Figure 4-1, calculate gpm of water flow if the heating coil load is 243,810 Btuh and TD is 20°F (200°F EWT-180°F LWT).

$$\text{Btuh} = \text{gpm} \times 500 \times \text{TD}$$

Where:

Btuh = Btu per hour

gpm = volume of water flow, gallons per minute

500 = constant (60 min/hr tant (water flow, gallons per)

TD= temperature difference of the water entering (EWT) and leaving (LWT) the coil. Then:

$$\text{gpm} = \text{Btuh} \div (500 \times \text{TD})$$

$$gpm = 243,810 \div (500 \times 20)$$

Answer:

24.4 gpm of water flows through the heating coil.

Now calculate the air TD across the heating coil if:

LAT - leaving air temperature coil is 105°F

EAT - entering air temperature coil is 62°F. EAT is MAT.

RAT - return or room air temperature is 70°F

OAT - outside air temperature is 30°F

198,450 Btuh is the Sensible Room Heating Load.

The math is:

$$198,450 = 5,250 \text{ cfm SA} \times 1.08 \times 35 \text{ TD } (105 - 70)$$

243,810 Btuh is the Sensible Coil Heating Load.

The difference of 45,360Btuh (243,810-198,450) is the additional heat required for the outside air.

The math is:

$$45,360 = 1,050 \text{ cfm OA} \times 1.08 \times 40 \text{ TD } (70 - 30)$$

Then:

$$TD = Btuh \div (1.08 \times cfm)$$

$$TD = 243,810 \div (1.08 \times 5,250)$$

Answer:

$$TD = 43°F$$

Then:

62°F EAT + 43°F TD = 105°F LAT

243,810 Btuh is the Coil Sensible Heating Load.

The math is

243,810 = 5,250 cfm × 1.08 × 43 TD (105 − 62)

The mixed air temperature (MAT also called EAT) was calculated using this equation:

MAT = (% OA × OAT) + (% RA × RAT)

Where:

MAT = mixed air temperature

OAT = outside air temperature

RAT = return air temperature, also called room air temperature

Then:

MAT = (20% × 30°F) + (80% × 70°F)

MAT = (6) + (56)

Answer:

MAT = 62°F

Unit 9 Ventilation Systems

In occupied buildings carbon dioxide, human odors and other contaminants such as volatile organic compounds (VOC) or odors and particles from machinery and other process functions need to be continuously removed or unhealthy conditions will result. Ventilation is the process of supplying "fresh" outside air to occupied buildings in the proper amount to offset the contaminants produced by people and equipment.

Today, local building codes, association and testing organization guidelines, and government or company protocols stipulate the amount of ventilation required for buildings and work environments. Ventilation guidelines in the USA are approximately 15 to 25cfm (cubic feet per minute) of air volume per person of outside air (OA). Ventilation air may also be required as additional or "make-up" air (MUA) for kitchen exhaust systems. Maintaining room or conditioned space pressurization (typically +0.03 to +0.05 inches of water gauge) in commercial and industrial buildings is part of proper ventilation.

9.1 Basic Concepts and Terminology

Outdoor air that flows through a building is often used to dilute and remove indoor air contaminants. However, the energy required to condition this outdoor air can be a significant portion of the total space-conditioning load. The magnitude of outdoor airflow into the building must be known for proper sizing of the HVACR equipment and evaluation of energy consumption. For buildings without mechanical cooling and dehumidification, proper ventilation and infiltration airflows are important for providing comfort for occupants. ASHRAE Standard 55 specifies conditions under which 80% or more of the occupants in a space will find it thermally acceptable. Additionally, airflow into buildings and between zones affects fires and the movement of smoke. Smoke management is addressed in Chapter 53 of the 2011*ASHRAE Handbook—HVAC Applications*.

Ventilation and Infiltration

Air exchange of outdoor air with air already in a building can be divided into two broad classifications: ventilation and infiltration. Ventilation is intentional introduction of air from the outdoors into a building; it is further subdivided into natural and mechanical ventilation. Natural ventilation is the flow of air through open windows, doors, grilles, and other planned building envelope penetrations, and it is driven by natural and/or artificially produced pressure differentials. Mechanical (or forced) ventilation, shown in Figure 9-1, is the intentional movement of air into and out of a building using fans and intake and exhaust vents.

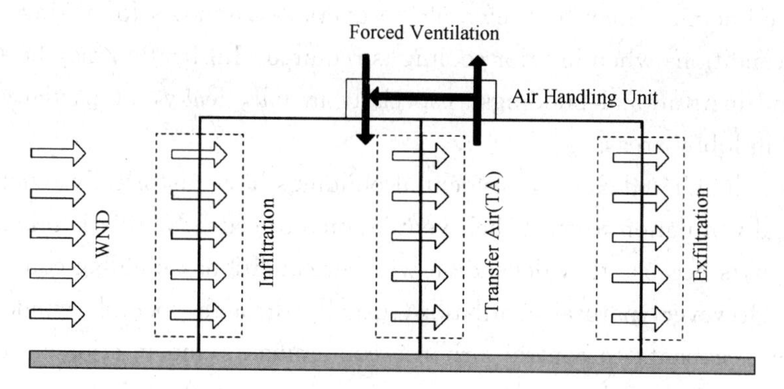

Figure 9-1 Two-space building with mechanical ventilation, infiltration, and exfiltration

Infiltration is the flow of outdoor air into a building through cracks and other unintentional openings and through the normal use of exterior doors for entrance and egress. Infiltration is also known as air leakage into a building. Exfiltration, depicted in Figure 9-1, is leakage of indoor air out of a building through similar types of openings. Like natural ventilation, infiltration and exfiltration are driven by natural and/or artificial pressure differences. These forces are discussed in detail in the section on Driving Mechanisms for Ventilation and Infiltration. Transfer air is air that moves from one interior space to another, either intentionally or not.

Ventilation and infiltration differ significantly in how they affect energy consumption, air quality, and thermal comfort, and they can each vary with weather conditions, building operation, and use. Although one mode may be expected to dominate in a particular building, all must be considered in the proper design and operation of an HVACR system.

Ventilation Air

Ventilation air is air used to provide acceptable indoor air quality. It may be composed of mechanical or natural ventilation, infiltration, suitably treated recirculated air, transfer air, or an appropriate combination, although the allowable means of providing ventilation air varies in standards and guidelines.

Modern commercial and institutional buildings normally have mechanical ventilation and are usually pressurized somewhat to reduce or eliminate infiltration. Mechanical ventilation has the greatest potential for control of air exchange when the system is properly designed, installed, and operated; it should provide acceptable indoor air quality and thermal comfort when ASHRAE *Standard* 55 and 62.1 requirements are followed. Mechanical ventilation equipment and systems are described in Chapters 1, 4, and 10 of the 2012 *ASHRAE Handbook—HVAC Systems and Equipment*.

In commercial and institutional buildings, natural ventilation (e.g., through operable windows) may not be desirable from the point of view of energy conservation and comfort.

In commercial and institutional buildings with mechanical cooling and ventilation, an air- or water-side economizer may be preferable to operable windows for taking advantage of cool outdoor conditions when interior cooling is required. Infiltration may be significant in commercial and institutional buildings, especially in tall, leaky, or partially pressurized buildings and in lobby areas.

In most of the United States, residential buildings have historically relied on infiltration and natural ventilation to meet their ventilation air needs. Neither is reliable for ventilation air purposes because they depend on weather conditions, building construction, and maintenance. However, natural ventilation, usually through operable windows, is more likely to allow occupants to control airborne contaminants and interior air temperature, but it can have a substantial energy cost if used while the residence's heating or cooling equipment is operating.

In place of operable windows, small exhaust fans should be provided for localized venting in residential spaces, such as kitchens and bathrooms. Not all local building codes require that the exhaust be vented to the outside. Instead, the code may allow the air to be treated and returned to the space or to be discharged to an attic space. Poor maintenance of these treatment devices can make non-ducted vents ineffective for ventilation purposes. Condensation in attics should be avoided. In northern Europe and in Canada, some building codes require general mechanical ventilation in residences, and heat recovery heat exchangers are popular for reducing energy consumption. Low-rise residential buildings with low rates of infiltration and natural ventilation, including most new buildings, require mechanical ventilation at rates given in *ASHRAE Standard* 62. 2.

Outdoor Air Fraction

The outdoor airflow introduced to a building or zone by an air-handling unit can also be described by the **outdoor** air fraction X_{oa}, which is the ratio of the volumetric flow rate of outdoor air brought in by the air handler to the total supply airflow rate:

$$X_{oa} = \frac{Q_{oa}}{Q_{sa}} = \frac{Q_{oa}}{Q_{ma}} = \frac{Q_{oa}}{Q_{oa} + Q_{ca}} \tag{9-1}$$

When expressed as a percentage, the outdoor air fraction is called the percent outdoor air. The design outdoor airflow rate for a building's or zone's ventilation system is found by applying the requirements of ASHRAE *Standard* 62. 1 to that specific building. The supply airflow rate is that required to meet the thermal load. The outdoor air fraction and percent outdoor air then describe the degree of recirculation, where a low value indicates a high rate of recirculation, and a high value shows little recirculation. Conventional all-air air-handling systems for commercial and institutional buildings have approximately 10 to 40% outdoor air.

100% outdoor air means no recirculation of return air through the air-handling system. Instead, all the supply air is treated outdoor air, also known as makeup air (KA), and all return air is discharged directly to the outdoors as relief air (LA), via separate or centralized exhaust fans. An air-handling unit that provides 100% outdoor air to offset air that is exhausted is typically called a makeup air unit (MAU).

Air Exchange Rate

The **air exchange** (or **change**) **rate**/compares airflow to volume and is

$$I=\frac{Q}{V} \tag{9-2a}$$

where Q = volumetric flow rate of air into space, m^3/s

V=interior volume of space, m^3

The air exchange rate has units of 1/time, usually h^{-1}. When the time unit is hours, the air exchange rate is also called **air changes per hour (ACH).** The air exchange rate may be defined for several different situations. For example, the air exchange rate for an entire building or thermal zone served by an air-handling unit compares the amount of outdoor air brought into the building or zone to the total interior volume. This **nominal air exchange rate** I_N is

$$I_N=\frac{Q_{oa}}{V} \tag{9-2b}$$

where Q_{oa} is the outdoor airflow rate including ventilation and infiltration. The nominal air exchange rate describes the outdoor air ventilation rate entering a building or zone. It does not describe recirculation or the distribution of the ventilation air to each space within a building or zone.

For a particular space, the space air exchange rate I_s compares the supply airflow rate Q_{sa} to the volume of that space:

$$I_S=\frac{Q_{sa}}{V} \tag{9-3}$$

The space air exchange rate for a particular space or zone includes recirculated as well as outdoor air in the supply air, and it is used frequently in the evaluation of supply air diffuser performance and space air mixing.

Age of Air

The age of air θ_{age} (Sandberg 1981) is the length of time t that some quantity of outdoor air has been in a building, zone, or space. The "youngest" air is at the point where outdoor air enters the building by mechanical or natural ventilation or through infiltration (Grieve 1989). The "oldest" air may be at some location in the build-ing or in the exhaust air. When the characteristics of the air distribution system are varied, age of air is inversely correlated with quality of outdoor air delivery. Units are of time, usually in seconds or minutes, so it is not a true efficiency or effectiveness measure. The age of air concept, however, has gained wide acceptance in Europe and is used increasingly in North America.

The age of air can be evaluated for existing buildings using tracer gas methods. Using either the decay (step-down) or growth (step-up) tracer gas method, the zone average or nominal age of air θ_{age}, N can be determined by taking concentration measurements in the exhaust air. The local age of air θ_{age}, L is evaluated through tracer gas measurements at any desired point in a space, such as at a worker's desk. When time-dependent data of tracer gas concentration are available, the age of air can be calculated from

$$\theta_{age} = \int_{t=0}^{\infty} \frac{C_{in} - C}{C_{in} - C_o} dt \tag{9-4}$$

where C_{in} is the concentration of tracer gas being injected. Because evaluation of the age of air requires integration to infinite time, an exponential tail is usually added to the known concentration data (Farrington et al. 1990).

Indoor Air Quality

Simply defined, indoor air quality (IAQ) is the state of the air contained within an enclosed building. The air may be conditioned or unconditioned, and it may be acceptable or objectionable to the occupants. To achieve acceptable IAQ, the air's temperature, humidity, and velocity may need to be conditioned and pollutants must be removed. Often, when people talk about IAQ they are talking about identifying any source of pollutants or contaminants and taking appropriate actions to remove, control, or prevent their amplification. Amplification in IAQ terms means the contaminant's ability to grow and spread.

The sick building syndrome (SBS) received public attention from 1970s after the energy crisis as a result of a tighter building and a reduced amount of outdoor ventilation air. Since Americans are spending more and more of their time indoors, they need a comfortable and healthy indoor environment and an acceptable indoor air quality. In an unhealthy building environment, uncomfortable employees do not perform well and productivity declines. Worker illness due to a poor indoor environment elevates absenteeism. A significant issue that is facing building owners, operating mangers, architects, consulting engineers, and contractors today is the possibility of legal suits filed by occupants or owners who feel that their health has been damaged by poor indoor air quality. In the 1900s, indoor air quality (IAQ) has become one of the primary concerns in air conditioning (HVACR) system design, manufacturing, installation, and operation because of the following:

- IAQ is closely related to the health of the occupants inside a building, whether a building is a healthy building or a sick building.
- IAQ and thermal control (zone temperature and relative humidity control) indicate primarily the quality of the indoor environment in a building.
- IAQ and thermal control represent mainly the functional performance of an air conditioning (HVACR) system.

"Indoor air quality" has rapidly advanced from a cutting—edge buzzword to an important consideration in building construction and operation. IAQ is now widely acknowledged and accepted as being important in the marketplace. However, the perception of what constitutes good IAQ and the associated improvements needed to achieve it remain an interesting topic of discussion and debate by those in the industry. For some, improving IAQ has meant simply upgrading air filters or the addition of an advanced air cleaner. For others, it has involved installing humidification or dehumidification systems, coil and drain pan cleanings, or duct cleanings, or even installing heat-recovery ventilation systems. Still to others it means looking at the entire building as a system and looking beyond just the HVACR systems to truly comprehend what is occurring in a building and why it is happening. It involves getting into the science that is affecting the air from a chemical and biological perspective as well as the physics of the air. To truly make a difference and solve IAQ issues properly, a technician needs to understand how a building and its systems interact. HVACR technicians need to understand that the most effective way to address IAQ is by using techniques that eliminate or reduce the source of the pollutants. It will sometimes take multiple strategies working together to fully address complex IAQ issues.

Outdoor air requirements for acceptable IAQ have long been debated, and different rationales have produced radically different ventilation standards (Grimsrud and Teichman 1989; Janssen 1989; Klauss et al. 1970; Yaglou et al. 1936; Yaglou and Witheridge 1937). Historically, the major considerations have included the amount of outdoor air required to control moisture, carbon dioxide (CO_2), odors, and tobacco smoke generated by occupants. These considerations have led to prescriptions of a minimum rate of outdoor air supply per occupant. More recently, a major concern has been maintaining acceptable indoor concentrations of various additional pollutants that are not generated primarily by occupants. Engineering experience and field studies indicate that an outdoor air supply of about 10 L/s per person is very likely to provide acceptable perceived indoor air quality in office spaces, whereas lower rates may lead to increased sick building syndrome symptoms (Apte et al. 2000; Mendell 1993; Seppanen et al. 1999).

Indoor pollutant concentrations depend on the strength of pollutant sources and the total rate of pollutant removal. Pollutant sources include outdoor air; indoor sources such as occupants, furnishings, and appliances; dirty ventilation system ducts and filters; soil adjacent to the building; and building materials themselves, especially when new. Pollutant removal processes include dilution with outdoor air, local exhaust ventilation, deposition on surfaces, chemical reactions, and air-cleaning processes. If (1) general building ventilation is the only significant pollutant removal process, (2) indoor air is thoroughly mixed, and (3) pollutant source strength and ventilation rate have been stable for a sufficient period, then the steady-state indoor pollutant concentration is given by

$$C_i = C_o + 10^6 S/Q_{oa} \tag{9-5}$$

where C_i = steady-state indoor concentration, ppm

C_o = outdoor concentration, ppm

S = total pollutant source strength, m^3/s

Q_{oa} = ventilation rate, m^3/s

Variation in pollutant source strengths (rather than variation in ventilation rate) is considered the largest cause of building-to-building variation in concentrations of pollutants that are not generated by occupants. Turk et al. (1989) found that a lack of correlation between average indoor respirable particle concentrations and whole-building outdoor ventilation rate indicated that source strength, high outdoor concentrations, building volume, and removal processes are important. Because pollutant source strengths are highly variable, maintaining minimum ventilation rates does not ensure acceptable indoor air quality in all situations. The lack of health-based concentration standards for many indoor air pollutants, primarily because of the lack of health data, makes the specification of minimum ventilation rates difficult.

In cases of high contaminant source strengths, such as with indoor sanding, spray painting, or smoking, impractically high rates of dilution ventilation are required to control contaminant levels, and other methods of control are more effective. Removal or reduction of contaminant sources is the most effective means of control. Controlling a localized source by means of local exhaust, such as range hoods or bathroom exhaust fans, as well as filtration and absorption, may also be effective [e. g. , Rock (2006)].

Particles can be removed with various types of air filters. Gaseous contaminants with higher relative molecular mass can be controlled with activated carbon or alumina pellets impregnated with a substance such as potassium permanganate. Chapter 29 of the 2012 *ASHRAE Handbook—HVAC Systems and Equipment* has information on air cleaning.

9.2 Natural Ventilation

Natural ventilation is the flow of outdoor air caused by wind and thermal pressures through intentional openings in the building's shell. Under some circumstances, it can effectively control both temperature and contaminants in mild climates, but it is not considered practical in hot and humid climates or in cold climates. Temperature control by natural ventilation is often the only means of providing cooling when mechanical air conditioning is not available. The arrangement, location, and control of ventilation openings should combine the driving forces of wind and temperature to achieve a desired ventilation rate and good distribution of ventilation air through the building. However, intentional openings cannot always guarantee adequate temperature and humidity control or indoor air quality because of the dependence on natural (wind and stack) effects to drive the flow (Wilson and Walker 1992). Using night ventilation and the building's thermal mass effect may be effective for reducing conventional cooling energy consumption in some buildings and climates if moisture condensation can be controlled. Axley (2001a) and the Chartered Institute of Building Services Engineers (CIBSE 2005) review natural ventilation in commercial buildings, including potential advantages and problems, natural ventilation components and system designs, and recommended design and analysis approaches.

Natural Ventilation Openings

Natural ventilation openings include (1) windows, doors, dormer (monitor) openings, and skylights; (2) roof ventilators; (3) stacks; and (4) specially designed inlet or outlet openings.

Windows transmit light and provide ventilation when open. They may open by sliding vertically or horizontally; by tilting on horizontal pivots at or near the center; or by swinging on pivots at the top, bottom, or side. The type of pivoting used is important for weather protection and affects airflow rate.

Roof ventilators provide a weather-resistant air outlet. Capacity is determined by the ventilator's location on the roof; the resistance to airflow of the ventilator and its ductwork; the ventilator's ability to use kinetic wind energy to induce flow by centrifugal or ejector action; and the height of the draft.

Natural-draft or gravity roof ventilators can be stationary, pivoting, oscillating, or rotating. Selection criteria include ruggedness, corrosion resistance, stormproofing features, dampers and operating mechanisms, noise, cost, and maintenance. Natural ventilators can be supplemented with power-driven supply fans; the motors need only be energized when the natural exhaust capacity is too low. Gravity ventilator dampers can be manual or controlled by thermostat or wind velocity.

A natural-draft roof ventilator should be positioned so that it receives full, unrestricted wind. Turbulence created by surrounding obstructions, including higher adjacent buildings, impairs a ventilator's ejector action. Inlets can be conical or bell-mouthed to increase their flow coefficients. The opening area at any inlet should be increased if screens, grilles, or other structural members cause flow resistance. Building air inlets at lower levels should be larger than the combined throat areas of all roof ventilators.

Stacks or vertical flues should be located where wind can act on them from any direction. Without wind, stack effect alone removes air from the room with the inlets.

Ceiling Heights

In buildings that rely on natural ventilation for cooling, floor-to-ceiling heights are often increased well beyond the normal 2.5 to 3.2 m. Higher ceilings, as seen in buildings constructed before air conditioning was available, allow warm air and contaminants to rise above the occupied portions of rooms. Air is then exhausted from the ceiling zones, and cooler outdoor air is provided near the floors; a degree of floor-to-ceiling displacement airflow is thus desirable when using natural ventilation for cooling.

Airflow Through Large Intentional Openings

The relationship describing the airflow through a large intentional opening is based on the Bernoulli equation with steady, incompressible flow. The general form that includes stack, wind, and mechanical ventilation pressures across the opening is

$$Q = C_D A \sqrt{2\Delta p \rho} \tag{9-6}$$

where Q = airflow rate, m^3/s

C_D = discharge coefficient for opening, dimensionless

A = cross-sectional area of opening, m^2

ρ = air density, kg/m^3

Δp = pressure difference across opening, Pa

The discharge coefficient C_D is a dimensionless number that de-pends on the geometry of the opening and the Reynolds number of the flow.

Flow Caused by Wind Only

Aspects of wind that affect the ventilation rate include average speed, prevailing direction, seasonal and daily variation in speed and direction, and local obstructions such as nearby buildings, hills, trees, and shrubbery. Liddament (1988) reviewed the relevance of wind pressure as a driving mechanism. A multiflow path simulation model was developed and used to illustrate the effects of wind on air exchange rate.

Wind speeds may be lower in summer than in winter; directional frequency is also a function of season. Natural ventilation systems are often designed for wind speeds of one-half the seasonal average. Equation (9-7) shows the rate of air forced through ventilation inlet openings by wind or determines the proper size of openings to produce given airflow rates:

$$Q = C_v A U \tag{9-7}$$

where Q = airflow rate, m^3/s

C_v = effectiveness of openings (C_v is assumed to be 0.5 to 0.6 for perpendicular winds and 0.25 to 0.35 for diagonal winds)

A = free area of inlet openings, m^2

U = wind speed, m/s

Inlets should face directly into the prevailing wind. If they are not advantageously placed, flow will be less than that predicted by Equation (9-7); if inlets are unusually well placed, flow will be slightly more. Desirable outlet locations are (1) on the leeward side of the building directly opposite the inlet; (2) on the roof, in the low-pressure area caused by a flow discontinuity of the wind; (3) on the side adjacent to the windward face where low-pressure areas occur; (4) in a dormer on the leeward side; (5) in roof ventilators; or (6) by stacks. Unit 10 gives a general description of the wind pressure distribution on a building. Inlets should be placed in exterior high-pressure regions; outlets should be placed in exterior low-pressure regions.

Flow Caused by Thermal Forces Only

If building internal resistance is not significant, flow caused by stack effect can be expressed by

$$Q = C_D A \sqrt{2g\Delta H_{NPL}(T_i - T_o)/T_i} \tag{9-8}$$

where Q = airflow rate, m^3/s

C_D = discharge coefficient for opening

ΔH_{NPL} = height from midpoint of lower opening to NPL, m

T_i = indoor temperature, K

T_o = outdoor temperature, K

Equation (9-8) applies when $T_i > T_o$. If $T_i < T_o$, replace T_i in the denominator with T_o, and replace $(T_i - T_o)$ in the numerator with $(T_o - T_i)$. An average temperature should be used for T_i if there is thermal stratification. If the building has more than one opening, the outlet and inlet areas are considered equal. The discharge coefficient C_D accounts for all viscous effects such as surface drag and interfacial mixing

Estimation of ΔH_{NPL} is difficult for naturally ventilated build-ings. If one window or door represents a large fraction (approxi-mately 90%) of the total opening area in the envelope, then the NPL is at the mid-height of that aperture, and ΔH_{NPL} equals one-half the height of the aperture. For this condition, flow through the opening is bidirectional (i. e., air from the warmer side flows through the top of the opening, and air from the colder side flows through the bottom). Interfacial mixing occurs across the counterflow interface, and the orifice coefficient can be calculated according to the following equation (Kiel and Wilson 1986)

$$C_D = 0.40 + 0.0045 \mid T_i - T_o \mid \tag{9-9}$$

If enough other openings are available, airflow through the opening will be unidirectional, and mixing cannot occur. A discharge coefficient of $C_D = 0.65$ should then be used. Additional information on stack-driven airflows for natural ventilation can be found in Foster and Down (1987).

Greatest flow per unit area of openings is obtained when inlet and outlet areas are equal; Equation (9-8) and (9-9) are based on this equality. Increasing the outlet area over inlet area (or vice versa) increases airflow but not in proportion to the added area. When openings are unequal, use the smaller area in Equation(9-8) and add the increase as determined from Figure 9-2.

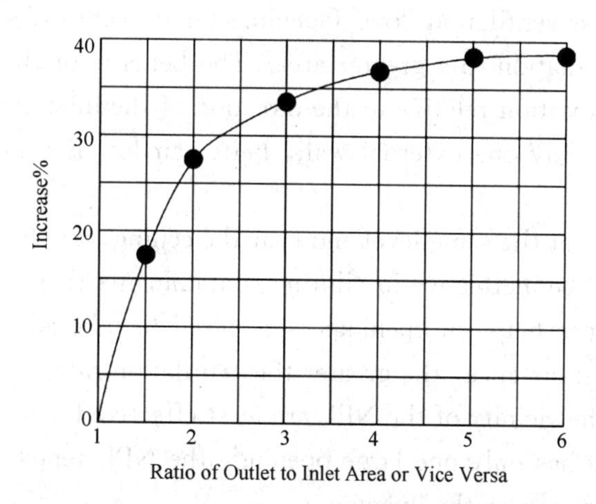

Figure 9-2 Increase in flow caused by excess area of one opening over the other.

Natural Ventilation Guidelines

Several general guidelines should be observed in designing for natural ventilation. Some of these may conflict with other climate-responsive strategies (such as using orientation and shading devices to minimize solar gain), with building codes that encourage compartmentalization to restrict fire and smoke movement, or with other design considerations.

System selection

• In hot, humid climates, use mechanical cooling. If mechanical cooling is not available, air velocities should be maximized in the occupied zones of rooms.

• In hot, arid climates, consider evaporative cooling. Airflow throughout the building should be maximized for structural cooling, particularly at night when the outdoor air temperature is low.

Building and surroundings characteristics

• Topography, landscaping, and surrounding buildings should be used to redirect airflow and give maximum exposure to breezes. Vegetation can funnel breezes and avoid wind dams, which reduce the driving pressure differential around the building. Site objects should not obstruct inlet openings.

• The building should be shaped to expose maximum shell openings to breezes.

• Architectural elements such as wing walls, parapets, and overhangs should be used to promote airflow into the building interior.

• The long façade of the building and the majority of door and window openings should be oriented with respect to prevailing summer breezes. If there is no prevailing direction, openings should be sufficient to provide ventilation regardless of wind direction.

Opening locations

• Windows should be located in opposing pressure zones. Two openings on opposite sides of a space increase ventilation flow. Openings on adjacent sides force air to change direction, providing ventilation to a greater area. The benefits of the window arrangement depend on the outlet location relative to the direction of the inlet airstream.

• If a room has only one external wall, better airflow is achieved with two widely spaced windows.

• If openings are at the same level and near the ceiling, much of the flow may bypass the occupied level and be ineffective in diluting contaminants there.

• Vertical distance between openings is required to take advantage of stack effect; the greater the vertical distance, the greater the ventilation rate.

• Openings in the vicinity of the NPL are least effective for ther-mally induced ventilation. If the building has only one large opening, the NPL tends to move to that level, which reduces pressure across the opening.

Opening characteristics

• Greatest flow per unit area of total opening is obtained by inlet and outlet openings of nearly equal areas. An inlet window smaller than the outlet creates higher inlet velocities. An outlet smaller than the inlet creates lower but more uniform airspeed through the room.

• Openings with areas much larger than calculated are sometimes desirable when anticipating increased occupancy or very hot weather.

• Horizontal windows are generally better than square or vertical windows. They produce more airflow over a wider range of wind directions and are most beneficial in locations where prevailing wind patterns shift.

• Window openings should be accessible to and operable by occupants, unless fully automated. For secondary fire egress, operable windows may be required.

• Inlet openings should not be obstructed by indoor partitions. Partitions can be placed to split and redirect airflow but should not restrict flow between the building's inlets and outlets. Vertical airshafts or open staircases can be used to increase and take advantage of stack effects. However, enclosed staircases intended for evacuation during a fire should not be used for ventilation.

9.3 Industrial Ventilation

Industrial environments require ventilation to reduce exposure to excess heat and contaminants that are generated in the workplace; in some situations, cooling may also be required. Ventilation is primarily used to control excess heat, odors, and hazardous particulate and chemical contaminants. These could affect workers' health and safety or, in some cases, become combustible or flammable when allowed to accumulate above their minimum explosible concentration (MEC) or lower flammable limit (LFL) [also called the lower explosive limit (LEL)] (Cashdollar 2009). Excess heat and contaminants can best be controlled by using local exhaust systems whenever possible. Local exhaust systems capture heated air and contaminants at their source and may require lower airflows than general (dilution) ventilation.

General ventilation can be provided by mechanical (fan) systems, by natural draft, or by a combination of the two. Combination systems could include mechanically driven (fan-driven) supply air with air pressure relief through louvers or other types of vents, and mechanical exhaust with air replacement inlet louvers and/or doors.

Mechanical (fan-driven) supply systems provide the best control and the most comfortable and uniform environment, especially when there are extremes in local climatic conditions. The systems typically consist of an inlet section, filtration section, heating and/or cooling equipment, fans, ductwork, and air diffusers for distributing air in the workplace. When toxic gases or vapors are not present and there are no aerosol contaminants associated with adverse health effects, air cleaned in the general exhaust system or in packaged air filtration units can berecirculated via a return duct. When applied appropriately, air recirculation can be a major contributor to a sustainable industrial ventilation design and may reduce heating and cooling costs.

In addition, regardless of the method selected, any positive ventilation into an industrial space should be from a source that will be essentially free of any contaminants under both normal and abnormal conditions in the surrounding atmosphere. In many cases, this may require a sealed intake stack or ductwork, as opposed to a perimeter wall hood or other air intake device, wherein the source of intake should be from a point well above or beyond the veil of the hazardous space that may surround a ventilated space.

A general exhaust system, which removes air contaminated by gases, vapors, or particulates not captured by local exhausts, usually consists of one or more fans, plus inlets, ductwork, and air cleaners or filters. After air passes through the filters, it is either discharged outside or partially recirculated to the building workplace. The air filter system's cleaning efficiency should conform to environmental regulations and depends on factors such as building location, background contaminant concentrations in the atmosphere, type and toxicity of contaminants, and height and velocity of the building exhaust discharge.

Many industrial ventilation systems must handle simultaneous exposures to heat and hazardous substances. In these cases, the required ventilation can be provided by a combination of local exhaust, general ventilation air supply, and general exhaust systems. The ventilation engineer must carefully analyze supply and exhaust air requirements to determine the worst case. For example, air supply makeup for hood exhaust may be insufficient to control heat exposure. It is also important to consider seasonal climatic effects on ventilation system performance, especially for natural ventilation systems.

Most importantly, if the hazardous substances are ignitable gases or dusts, all electrical components of the ventilation system should be rated for the proper electrical classification in the absence of any ventilation, regardless of their locations in the ventilation system.

In specifying acceptable chemical contaminant and heat exposure levels, the industrial hygienist or industrial hygiene engineer must consult the appropriate occupational exposure limits that apply as well as any governing standards and guidelines. The legislated limits for the maximum airborne concentration of chemical substances to which a worker may be exposed are listed as (1) maximum average exposures to which a worker may be exposed over a given work day (generally assumes an 8 to 10 h work day and a traditional 40 h work week); (2) short-term exposure limits, which are the maximum average airborne concentration to which a worker may be exposed over any 15 min period; and (3) ceiling limits, which are the maximum airborne concentration to which a worker may be exposed at any time. However, occupational exposure limits for cold, heat, and contaminants are not lines of demarcation between safe and unsafe exposures. Rather, they represent conditions to which it is believed nearly all workers may be exposed day after day without adverse and/or long-term effects. Because a small percentage of workers may be affected by occupational exposure below the regulated limits, it is prudent to design for exposure levelswell below the limits.

In the case of exposure to hazardous chemicals, the number of contaminant sources, their generation rates, and the effectiveness of exhaust hoods may not be known. Consequently, the ventilation engineer must rely on industrial hygiene engineering practices when designing toxic and/or hazardous chemical controls. Close cooperation among the industrial hygienist, process engineer, and ventilation engineer is required.

In the case of exposure to flammable or ignitable chemicals, the specific gravity of the contaminant source(s), their concentration, and the rating of all electrical devices within the space, along with any source or point of excessive heat, must be carefully considered to prevent possible loss of life or severe injury. As with all hazardous chemicals, cooperation of knowledgeable experts, including electrical engineers, is required.

Ventilation Design Principles

Special Warning: Certain industrial spaces may contain flammable, combustible, and/or toxic concentrations of vapors or dusts under either normal or abnormal conditions. In spaces such as these, there are life safety issues that this chapter may not completely address. Special precautions must be taken in accordance with requirements of recognized authorities such as the National Fire Protection Association (NFPA), the Occupational Safety and Health Administration (OSHA), and the American National Standards Institute (ANSI). In all situations, engineers, designers, and installers who encounter conflicting codes and standards must defer to the code or standard that best addresses and safeguards life safety.

General Ventilation

General ventilation supplies and/or exhausts air to provide heat relief, dilute contaminants to an acceptable level, and replace exhaust air. Ventilation can be provided by natural or mechanical supply and/or exhaust systems. Industrial areas must comply with ASHRAE *Standard* 62. 1-2013 and other standards as required (e. g. , by NFPA). Outdoor air is unacceptable for ventilation if it is known to contain any contaminant at a concentration above that given in ASHRAE *Standard* 62. 1. If air is thought to contain any contaminant not listed in the standard, guidance on acceptable exposure levels should be obtained from relevant federal, state, provincial, or local jurisdictions. In addition to their role in controlling industrial contaminants, general ventilation rates must be sufficient to dilute the carbon dioxide produced by occupants.

For complex industrial ventilation problems, experimental scale models and computational fluid dynamics (CFD) models are often used in addition to field testing.

Makeup Air

When large volumes of air are exhausted to provide acceptable comfort and safety for personnel and acceptable conditions for process operations, this air must be replaced, either through intentional design strategy or through paths of least resistance. A safe and effective ventilation design should be strategic about the mechanism, locations, and physical parameters by which the makeup air enters the occupied space. Makeup air, consistently provided by good air distribution, allows more effective cooling in the summer and more efficient and effective heating in the winter. When makeup air design is not incorporated into the ventilation design scheme, it may lead to inefficient operation of local exhaust systems and/or combustion equipment and cross-drafts that affect occupant comfort and environmental control settings. Relying on windows or other air inlets that cannot function in year-round weather conditions is discouraged. Some factors to consider in makeup air design include the following:

• Makeup air must be sufficient to replace air being exhausted or consumed by combustion processes, local and general exhaust systems, or process equipment. (Large air compressors can consume a large amount of air and should be considered if air is drawn from within the building.)

• Makeup air systems should be designed to eliminate uncomfortable crossdrafts by properly arranging supply air outlets, and to prevent infiltration (through doors, windows, and similar openings) that may make hoods unsafe or ineffective, defeat environmental control, bring in or stir up dust, or adversely affect processes by cooling or disturbances. The design engineer needs to consider side drafts and other sources of air movement close to the capture area of a local exhaust hood. In industrial applications, it is common to see large fans blowing air onto workers positioned in front of the hood. This can render the local exhaust hood ineffective to the point that no protection is provided for the worker: Ahn et al. (2008), Caplan and Knutson(1977, 1978), and Tseng et al. (2010) found that air movement infront of laboratory hoods can cause contaminants to escape fromthe hood and into the operator's breathing zone. Hoods should belocated safe distances from doors and openable windows, supplyair diffusers, and areas of high personnel traffic (AIHA *Standard* Z9. 5; NFPA *Standard* 45).

• Makeup air should be obtained from a clean source with no more than trace amounts of any airborne contaminants or hazardous, ignitable substances. Supply air can be filtered, but infiltration air cannot. For transfer air use, see ASHRAE *Standard* 62. 1.

• Makeup air for spaces contaminated by toxic, ignitable, or combustible chemicals may have to be acquired through carefully sealed ductwork from an area know to be free of contamination and be supplied at sufficient rates, pressures, and mixing efficiencies to (1) remove all contamination, and (2) prevent infiltration of similar contaminants from surrounding areas or adjacent spaces.

• Makeup air should be used to control building pressure and airflow from space to space to (1) avoid positive or negative pressures that make it difficult or unsafe to open doors, (2) minimize drafts, and (3) prevent infiltration.

• Makeup air should be used to reduce contaminant concentration, to control temperature and humidity, and minimize undesirable air movement.

• Makeup air systems should be designed to recover heat and conserve energy (see the section on Energy Conservation, Recovery, and Sustainability).

For more information on potential adverse conditions caused by specific negative pressure levels in buildings, see ACGIH (2013) and Chapter 28 in the 2012 *ASHRAE Hand-*

book—HVAC Systemsand Equipment.

General Comfort and Dilution Ventilation

Effective air diffusion in ventilated rooms and the proper quantity of conditioned air are essential for creating an acceptable working environment, removing contaminants, and reducing installation and operating costs of a ventilation system. Ventilation systems must supply air at the proper velocity and temperature, with resulting contaminant concentrations within permissible occupational exposure limits (OELs). For the industrial environment, the most common objective is to provide tolerable (acceptable) working conditions rather than comfort (optimal) conditions.

General ventilation system design is based on the assumption that local exhaust ventilation, radiation shielding, and equipment insulation and encapsulation have been selected to minimize both heat load and contamination in the workplace (see the section on Heat Control). When work operations are generally restricted, such as with equipment operating stations or control booths, spot conditioning of the work environment with clean conditioned air (see the preceding section on Makeup Air) may further reduce the reliance on general ventilation for conditioning or contaminant dilution. In cold climates, infiltration and heat loss through the building envelope may need to be minimized by pressurizing buildings.

For more information on dilution ventilation, see ACGIH (2013).

Quantity of Supplied Air Sufficient

Sufficient air must be supplied to replace air exhausted by process ventilation and local exhausts, dilute Contaminants (gases, vapors, or airborne particles) not captured by local exhausts, prevent the entry of contaminants or hazardous (ignitable) substances from any surrounding atmosphere during ingress or egress, and provide the required thermal environment. The amount of supplied air should be the largest of the amounts needed for temperature control, dilution, and replacement.

Air Supply Methods

Air supply to industrial spaces can be by natural or mechanical ventilation systems. Although natural ventilation systems driven by gravity forces and/or wind effect are still widely used in industrial spaces (especially in hot premises in cold and moderate climates), they are inefficient in large buildings, may cause drafts, and may not solve air contamination problems because there is no practical filtration method available. Thus, most ventilation systems in industrial spaces are either mechanical (fan-driven) or a combination of mechanical supply with natural exhaust, using louvers or doors for air pressure relief (or for air replacement in exhaust systems).

The most common methods of air supply to industrial spaces are mixing, displacement, and localized.

Mixing Air Distribution. In mixing systems, air is normally supplied at velocities much greater than those acceptable in the occupied zone. Supply air temperature can be above, below, or equal to the air temperature in the occupied zone, depending on the heating/cooling load. The supply air diffuser jet mixes with room air by entrainment, which reduces air velocities and equalizes the air temperature. The occupied zone is ventilated either directly by the air jet or by reverse flow created by the jet. Properly selected and designed mixing air distribution creates relatively uniform air velocity, temperature, humidity, and air quality conditions in the occupied zone and over the room height. Note that supply systems should introduce air into the workspace in such a way as to not interfere with contaminant control systems such as ventilation hoods. If possible, ventilation hoods (e. g. , fume hoods) should have quiescent air conditions (~0. 5 m/s) at their face.

Displacement Ventilation Systems. Conditioned air that is slightly cooler than the desired room air temperature in the occupied zone is supplied from air outlets at low air velocities (~0. 5 m/s or less). Because of buoyancy, the cooler air spreads along the floor and floods the room's lower zone. Air close to the heat source is heated and rises upward as a convective air stream; in the upper zone, this stream spreads along the ceiling. The height of the lower zone depends on the air volume and temperature supplied to the occupied zone and on the amount of convective heat discharged by the sources.

Typically, outlets are located at or near the floor, and supply air is introduced directly into the occupied zone. In some applications (e. g. , in computer rooms or hot industrial buildings), air may be supplied to the occupied zone through a raised floor. Exhaust or air returns are located at or close to the ceiling or roof.

Displacement ventilation is common in European countries. It is an option when contaminants are released in combination with surplus heat, and contaminated air is warmer (more buoyant) than the surrounding air. It is not a good choice when air turbulence can interfere with convective conveyance of heat and contaminants. Further information on displacement air distribution systems can be found in Goodfellow and Tahti (2001).

Localized Ventilation. Air is supplied locally for occupied regions or a few permanent work areas as shown in Figure 9-3. Conditioned air is supplied toward the breathing zone of the occupants to create comfortable conditions and/or to reduce the concentration of pollutants. These zones may have air 5 to 10 times cleaner than the surrounding air. In localized ventilation systems, air is supplied through one of the following devices:

- Nozzles or grilles (e. g. , for spot cooling), specially designed low-velocity/ low-turbulence devices
- Perforated panels suspended on vertical duct drops and positioned close to the workstation

Local Area and Spot Cooling

In hot workplaces that have few work areas, it is likely impractical to maintain a comfortable environment in the entire space. However, environmentally controlled cabins, individual cooling, and spot cooling and extraction can improve working conditions in occupied areas. Certain applications require minimum distances from supplied air (and natural ventilation points) to ensure air flow in ventilated hoods and cabinets is not affected.

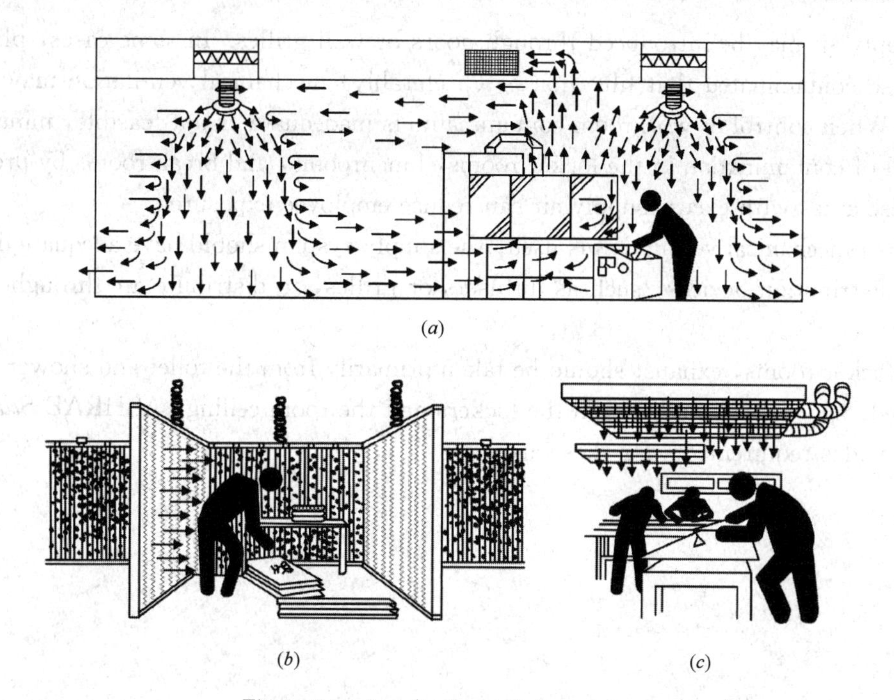

Figure 9-3 Localized ventilation systems
(*a*) Air showers; (*b*) Air oasis with horizontal air supply; (*c*) Air oasis with vertical air supply

Environmentally controlled cabins (e. g., operating cabs, pulpits, control rooms, enclosures) can not only provide thermal comfort, but when pressurized with a dedicated clean air supply (either dedicated source or through effective filtration), also can improve air quality in workers' occupational environments. There usually are significant economic benefits to properly designing, installing, and maintaining worker-protective environmental enclosures.

Spot cooling, probably the most popular method of improving the thermal environment, can be provided by radiation (changing mean radiant temperature), by convection (changing air velocity and/or air supply temperatures), or both. Spot-cooling equipment is fixed at the workstation, whereas in individual cooling, the worker wears the equipment.

Local exhaust ventilation (spot extraction) is another method to remove excess heat from a process or source of high temperature, and should be the first step considered for energy saving over spot cooling.

Locker Room, Toilet, and Shower Space Ventilation

Ventilation of locker rooms, toilets, and shower spaces is important in industrial facilities to remove odor and reduce humidity. In some industries, adequate control of work-room contamination requires prevention of both ingestion and inhalation routes of exposure, so adequate hygienic facilities, including appropriate ventilation, may be required in locker rooms, changing rooms, showers, lunchrooms, and break rooms. State, provincial, and local regulations should be consulted early in design.

Supply air may be introduced through doors or wall grilles. In some cases, plant air may be so contaminated that filtration or (preferably) mechanical ventilation may be required. When control of workroom contaminants is inadequate or not feasible, minimizing the level of contamination in the locker rooms, lunchrooms, and break rooms by pressurizing these areas with excess supply air can reduce employee exposure.

When mechanical ventilation is used, the supply system should have adequate ducting and air distribution devices, such as diffusers or grilles, to distribute air throughout the area.

In locker rooms, exhaust should be taken primarily from the toilet and shower spaces as needed, and the remainder from the lockers and the room ceiling. ASHRAE *Standard* 62.1 provides requirements for these areas.

Unit 10 Air Distribution

After the fan produces airflow the other air distribution components convey the air to and from occupied and non-occupied work and process spaces to heat, cool, ventilate and remove contaminates. These components include ductwork or duct, dampers, diverters, air valves, terminal boxes, supply outlets, and return and exhaust inlets. Air moves through the duct work because of a difference in pressures. Just as heat moves from a higher level to a lower level, so do fluids.

Fluids move from a higher pressure to a lower pressure. Air is a vapor and as such is a compressible fluid. Air moves through the duct system because the pressure on one side (discharge) of the fan is higher than on the inlet or suction side of the fan. The fan produces a pressure at the discharge of the fan that is higher than the pressure in the conditioned space, i. e., the pressure in the conditioned space is atmospheric pressure while the pressure at the fan discharge is greater than atmospheric pressure.

Duct systems are the distribution network for conditioned air to be moved throughout a building. Technicians need to understand duct systems and airflow to be able to troubleshoot and maintain HVACR systems. The most common materials used for duct construction include galvanized steel, fiberglass ductboard, and wire-helix flexible duct. Regardless of the materials used, a correctly installed duct system should last as long as the house, should not leak, and should be insulated well enough to prevent duct loss and gain to unconditioned space.

10.1 Air-Distribution System Components

Forced-air systems are used to distribute conditioned air in residential and small commercial buildings. Air is conditioned and distributed through the duct system throughout the building. The basic components of a forced-air system are the blower, the return-air ductwork carrying air to the blower, and the supply-air ductwork carrying air from the blower to the building.

Most systems have large boxes on both the return and supply ends of the blower called plenums. The plenum distributes the air to the ductwork attached to it. The duct design can be compared to a tree: the main ducts leaving the plenum are called trunk ducts, and the individual ducts running to each room are called branch ducts. The point where a branch duct comes off of a trunk duct is called a takeoff.

The duct openings to the conditioned space are covered by registers, diffusers, or grilles. The return-air openings are covered by return-air grilles. The supply-air openings are covered by supply-air registers. They direct the airflow into the room. These are sometimes called diffusers because they spread, or diffuse, the air.

Between the blower and the diffuser, the airstream must change direction and shape. This is accomplished with duct fittings. Turns are made with elbows, often called ells. A wye fitting is used to split one large duct into two smaller ducts. A change in the size of a rectangular duct is accomplished with a transition. A change in the size of a round duct is done with a reducer. When the register at the end of a round branch duct is rectangular, a fitting called a boot is used to allow the round duct to connect to the rectangular register.

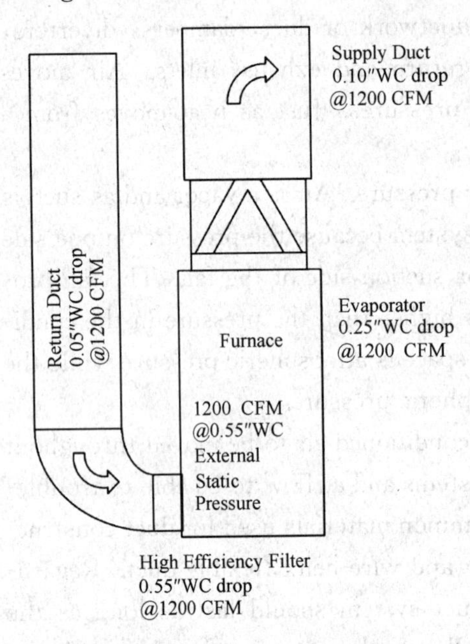

Figure 10-1 Basic components of a forced-air system.

The blower provides the pressure difference to move the air through the ductwork. The amount of air the blower can move and the amount of energy needed to move the air is controlled by the resistance to airflow from the ductwork and all the system components in the airstream. The duct offers resistance to airflow, creating a pressure drop as the air travels through it. Besides the ducts, every component that the air travels through adds pressure drop. This includes filters, humidifiers, heat exchangers, coils, registers, and grilles. The amount of pressure left for moving air through the ductwork is the difference between the amount of pressure the fan can produce and the amount of pressure drop from all these system components. Figure 10-1 shows the components of a forced-air system and the pressure drop created across each component.

Airflow in HVACR system is controlled by using dampers, diverters, air valves, terminal boxes, supply air outlets and return and exhaust air inlets.

10.2 Duct System Types

The four most common duct configurations are radial, reducing radial, extended plenum, and reducing extended plenum. Two less common types are the perimeter loop system and the central plenum system.

Radial Duct Systems

Radial duct systems are designed so that all or almost all of the duct runs originate at the central plenum. In some cases, a few of the duct runs may have wyes or duct triangles as a means of joining additional ducts to an initial run as shown in Figure 10-2. Radial systems are most frequently installed in attics, but they can also be installed in crawlspaces and basements. Radial systems are commonly used in small houses built on concrete slabs.

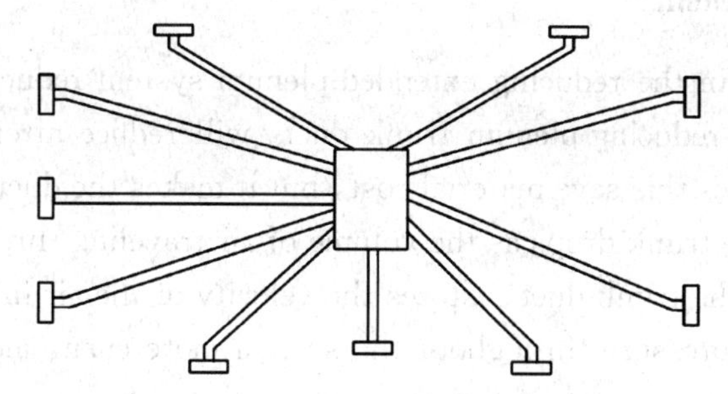

Figure 10-2 Radial duct system.

Reducing Radial Duct Systems

A reducing radial system uses several larger ducts leaving the main plenum that branch into smaller ducts as they get closer to their destination as shown in Figure 10-3. This reduces the connections at the plenum and reduces the overall amount of duct used.

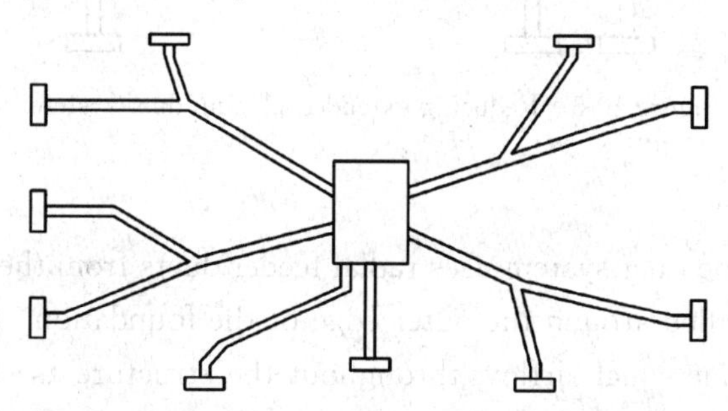

Figure 10-3 Reducing radial duct system.

Extended Plenum

The extended-plenum duct system uses a large trunk duct that travels the length of the building from the air handler. These systems are sometimes referred to as trunk duct systems as shown in Figure 10-4. The trunk duct is considered an extension of the plenum. The trunk ducts in extended plenum systems do not reduce in size as they travel across the structure. Extended plenum systems can be located in the crawlspace, attic, or basement.

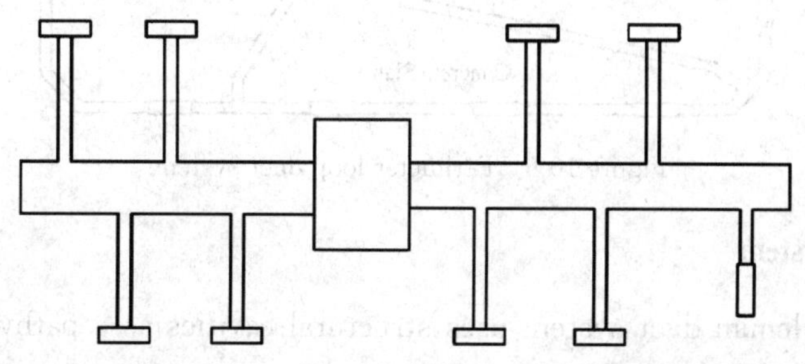

Figure 10-4 Extended plenum duct system.

Reducing Extended Plenum

The trunk ducts in the reducing extended-plenum system reduce in size as shown in Figure 10-5. Typical reducing-plenum trunk ducts will reduce after every three to four takeoffs. Not only does this save material cost, but it makes the duct system work better. The air velocity in the trunk drops as the volume of air traveling through the trunk drops. Reducing the size of the trunk duct restores the velocity of the air in the trunk duct. This helps keep the static pressure throughout the system more even, aiding in more even air distribution.

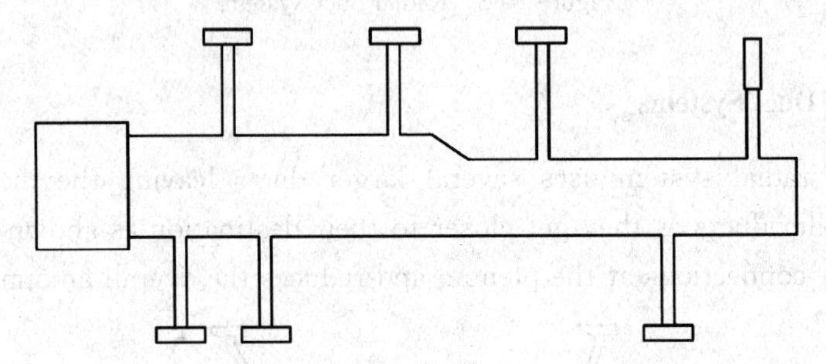

Figure 10-5 Reducing extended plenum duct system.

Perimeter Loop System

The perimeter-loop duct system uses radial feeder ducts from the blower that attach to one trunk that is installed around the outer edge of the foundation. Each supply is tapped off this trunk to provide equal airflow throughout the structure as shown in Figure 10-6. Perimeter-loop systems are normally installed in concrete foundations and have the added efficiency of heating the slab. The duct for a perimeter-loop system must be placed before the slab is poured. These systems are generally used in single-story commercial office buildings built on slabs.

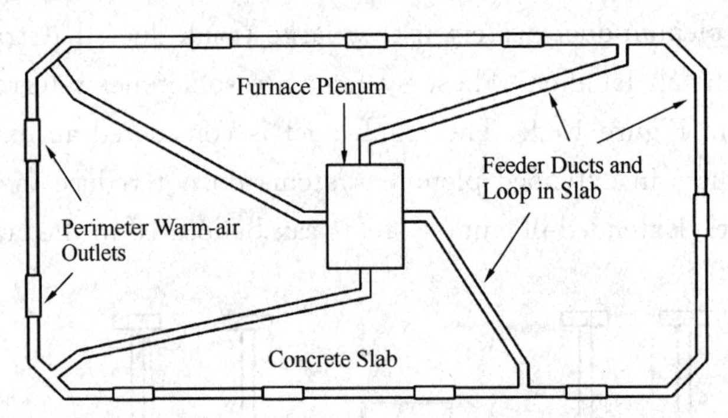

Figure 10-6 Perimeter loop duct system.

Central Plenum System

The central plenum duct system uses structural cavities as a pathway for supply- or return-air plenums. The basement, crawlspace, or space between floor joists can all serve as the plenum. Air is blown into the cavity, and holes are cut through the floor into the plenum cavity wherever a register is needed. To ensure proper operation of a plenum system, the cavity must be airtight and configured in a way that the airflow is not compromised.

This system is used because of its low cost. This system may not meet mechanical codes in some jurisdictions. Indoor air quality is a major concern due to the nature of these cavities. Moisture, mold, and odors are often hard to control.

10.3 Duct Materials and Insulation

Air ducts can be made from many different materials. The most common types of ductwork are galvanized-steel sheet metal, spiral metal, fiberglass ductboard, flexible duct, and fabric duct.

Galvanized Sheet Metal

Sheet metal can be fabricated into most any shape imaginable by a skilled sheet-metal worker. Galvanized sheet steel comes in large, flat sheets. Common sizes are 4 ft wide and 8-10 ft long, but other sizes are available depending upon the steel supplier. Sheet metal offers the least amount of resistance to airflow of any duct material because of its smooth surface.

Many localities adopt the metal duct standards developed by the Sheet Metal and Air Conditioning Contractor's National Association (SMACNA). The thickness of the metal is called its gauge. A guideline for selecting metal thickness is shown in Table 10-1.

Gauges recommended for sheet-metal ductwork. **Table10-1**

	Comfort Heating or Cooling			Comfort Heating Only
	Galvanized Steel		Approximate aluminum B&S gauge	
	Nominal thickness (in inches)	Equivalent galvanized sheet gauge no.		Minimum weight tin plate pounds per base box
Round ducts and enclosed rectangular ducts				
14 in or less	0.016	30	26	135
Over 14 in	0.019	28	24	—
Exposed rectangular ducts				
14 in or less	0.019	28	24	—
Over 14 in	0.022	26	23	—

For many years, galvanized sheet steel was used exclusively for air-conditioning ductwork because of its workability and durability. However, the material is expensive and costly to install, so other types of duct material have become popular. These include spiral metal duct, fiberglass ductboard, and flexible duct.

Spiral Metal Duct

Spiral duct is made from long strips of narrow metal and fabricated with spiral seams. Machines are available for making ducts on the job to fit required diameters and lengths. Spiral metal duct is used in commercial applications. It is frequently used where the ductwork will be exposed. It requires less support than other types of duct due to its inherent rigidity.

The UL 181 standard covers factory-made ducts and air connectors. All fiberduct and flexible duct should be UL approved and meet the UL 181 standard.

Fiberglass Ductboard

Fiberglass ductboard is a rigid material made of compressed fiberglass with an outer vapor barrier. It comes in 1-in, 1½-in, and 2-in thicknesses. Fiberglass ductboard has the advantage of being an inherently good insulator for both heat and sound. This reduces the duct losses and provides sound-absorbing qualities. Fiberglass ductboard is less expensive than metal, and fabricating ductboard duct is generally easier to work with than fabricating sheet metal duct. Ductboard is less durable than sheet metal duct and has a higher resistance to airflow than sheet metal because of its rougher interior surface.

Flexible Duct

The most common duct material used today in residential duct systems is undoubtedly flexible duct. Flexible duct has a spiral metal wire for support, a smooth plastic inner liner, an outer cover that serves as a vapor barrier, and fiberglass insulation sandwiched in between the inner liner and outer cover. The outer cover is typically vinyl or Mylar. Flexible duct comes in 25-ft lengths compressed in a box that is about 3 ft long. When opened it expands lengthwise into ducts.

It is very important to stretch flex duct before installing it. Leaving the duct slack can easily double the duct's resistance to airflow. Allowing extra length in the duct run is a particularly bad practice that may save time during installation but will significantly reduce the airflow through the duct.

Flex duct is very popular because it is the least expensive duct material and the easiest to install. Unfortunately, it is also the easiest to install incorrectly. Common problems with flex duct include tight radius turns and improper support. Many poorly installed flex duct jobs have given the material a bad name. However, it is possible to install a good duct system with flex duct by following the manufacturer's installation instructions. Flex duct is very quiet because its soft sides absorb sound. However, the soft, undulating sides also have the highest resistance to airflow of all of the most commonly used duct materials.

Duct Insulation

One major disadvantage of metal duct compared to ductboard or flex is that metal duct must be insulated. Round duct is insulated with duct wrap. The most common type of duct wrap is made of fiberglass with an outer vapor barrier of vinyl, aluminum foil, or aluminum foil reinforced with kraft-back paper. Duct wrap comes in 4-ft rolls. One side has 2 in of vapor barrier that is uninsulated. This edge laps over the edge of the insulation already on the duct, providing a continuous vapor barrier. When insulating duct, the duct wrap is pulled around the duct and lapped over itself. The insulation should be pulled tightly around the duct to eliminate gaps and sags in the insulation. But compressing insulation reduces its R value, so it should not be pulled so tightly that the fiberglass is flattened. The most common method for fastening the duct wrap is stapling it to itself. A clinch staple goes through both layers of insulation and turns back. The insulation should be stapled every 6 in for the entire length. The exposed outer edge is taped with a UL181A tape that matches the vapor barrier. When the job is done, no metal should be showing, no fiberglass should be showing, and the vapor barrier should cover the entire duct system.

The duct wrap should be cut in pieces that are large enough to wrap around the duct and have some overlap. For round duct, this is typically the duct circumference plus an adjustment for the thickness of the wrap plus some overlap. The adjustment is 9.5 in for 1.5-in-thick wrap, 12 in for 2-in-thick wrap, and 17 in for 3-in-thick wrap. Table 10-2 shows the size that duct wrap should be cut for different diameter ducts and different thickness of duct wrap. For example, 2-in-thick insulation for a 7-in-diameter duct should be 34-in wide. The same basic concept applies to rectangular duct, but the adjustments are not quite as large. The adjustments are 7 in for 1.5-in-thick insulation, 8 in for 2-in-thick insulation, and 11.5 in for 3-in-thick insulation.

Duct Insulation Dimensions (in inches) **Table 10-2**

Duct Dimensions		Insulation Thickness		
Duct Diameter	Duct Circumference	1.5 in (9.5-in Adjustment)	2 in (12-in Adjustment)	3 in (17-in Adjustment)
5	16	25.5	28	33
6	19	28.5	31	36
7	22	31.5	34	39
8	25	34.5	37	42
9	28	37.5	40	45
10	31.5	41	43.5	48.5
11	35	44.5	47	52
12	38	47.5	50	55
14	44	53.5	56	61
16	50	59.5	62	67

Duct Liner

Rectangular duct, plenums, and return-air boxes may be lined with insulation rather than wrapped. Duct liner is also fiberglass, but it does not have a vapor barrier. Duct liner is glued to the inside of the duct and connected with mechanical fasteners that compress the insulation. The fasteners should start within 4 in of the end of the duct and continue every 18 in for the length of the duct. A new row of fasteners should be installed every 12 in across the width the duct. Lining the duct makes a system quieter and is usually easier than wrapping the duct. A disadvantage of lining duct is increased resistance to air flow. One of the advantages of metal duct, low air resistance, is negated by lining it with fiberglass. Liner also tends to collect dust. After a period of years, the liner will end up looking like a dirty filter because the fiberglass grabs and holds tiny dust particles. When a liner is used, the duct must be made larger to compensate for the space the liner takes up. When using 1-in liner, the duct must be 2 in larger in both dimensions to allow for the thickness of the duct liner.

Bubble Wrap

Bubble wrap is a new form of duct insulation. It is similar to the bubble packing material with outer layers of either metalized Mylar or aluminum. Bubble wrap may be used as liner or as duct wrap. Bubble wrap is typically not applied directly to the duct. Spacers are applied to the duct and the bubble wrap is applied to the spacers. Installers like bubble wrap because it does not irritate their skin and cause itching like fiberglass does. However, it is a new product without the long track record of success that fiberglass has.

Unit 11 Water Distribution

After the pump produces water flow the water distribution components convey the water to heat and cool the occupied and non-occupied work and process spaces. Water systems are part of an air conditioning system and link the central plant, chiller/boiler, air-handling units (AHUs), and terminals. The components include: temperature, flow, volume, and pressure measuring stations, pipe systems, filtration, flow control, water flow balancing stations, pressure control components, air control components, and heat conversion equipment.

11.1 Hydronic Pipe Systems

HVACR hydronic pipe systems heat and cool with water. Pipe system can be classified in many ways including: open, closed, series loop direct return, reverse return, one-pipe, two-pipe, four-pipe, and primary or primary-secondary and combination as a pipe system may combine several of the piping arrangements mentioned. The pipe circuits provide heated or chilled water to coils in central air handling units, fan-coil units, ducts, terminal boxes, unit heaters, valence units and fin-tube radiation. Hydronic systems for central air handling units are typically four-pipe heating and cooling circuits.

Closed System and Open System

A closed pipe system is one in which there's no break in the piping circuit and the water is "closed" to the atmosphere. An open pipe system is one in which there's a break in the piping circuit and the water is "open" to the atmosphere, as shown in Figure 11-1.

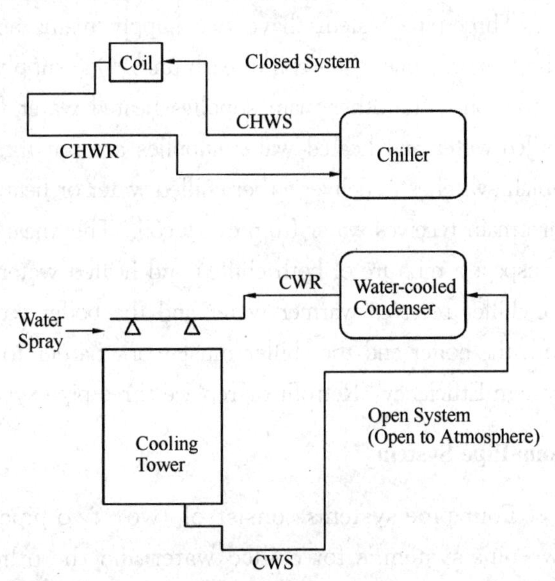

Figure 11-1 Closed pipe chilled water system and open pipe condenser water system. CHWR: Chilled Water System. CWS: Condenser Water System

An example of an open system is a water-cooled condenser and cooling tower. Examples of closed pipe systems are a chilled water system transporting water to and from the water chiller and to and from cooling coils, or a heated water system transporting water from the water boiler to heating coils and then back to the boiler.

One-Pipe System

One pipe systems have been used for heating in residences, small commercial buildings and industrial buildings. A one pipe system uses a single loop main distribution pipe. Each terminal coil is connected by a supply and return branch pipe to the main. System efficiency: a diverting tee should be installed in either the supply branch, return branch or sometimes both branches. Select diverting tees to create the proper amount of resistance in the main to direct water to the coil. If the diverting device is not installed, the water circulating in the main will tend to flow through the straight run of a normal tee and not be directed into the coil. This is because the coil has a higher pressure drop than the main. If this happens the coil is"starved" for water. Additionally, installing separate control valves and service valves in the branches will provide a better system. If the system has control valves and there are too many terminals, the coils farthest from the boiler may not receive water at a temperature that is high enough to maintain desired space temperatures.

Two-Pipe System

Two pipe systems have a supply pipe and a return pipe to each coil. There are separate automatic control valves and manual service valves for each water coil. For larger water systems, use two-pipe arrangements to maintain the water temperature to each coil equal to the boiler temperature.

Three-Pipe System

Three-pipe systems have two supply mains and one return main to each HVACR unit. There is only one water coil in each unit. One supply main provides chilled water from the chiller to the coil. The other main supplies heated water from the boiler to the coil in the unit. The chilled water and heated water supplies are not mixed. Each water coil has a three-way valve which switches to deliver either chilled water or heating water (but not both) to the coil. The return main receives water from every coil. This means that in some instances the return pipe may transport a mixture of both chilled and heated water. This results in a waste of energy because the chiller receives warmer water and the boiler receives cooler water than design. Therefore, both the boiler and the chiller must work harder to supply their proper discharge temperature. System Efficiency: Retrofit or replace three-pipe systems.

Four-Pipe System

Four-pipe systems consist of two, two-pipe systems as shown in Figure 11-2. One two-pipe system is for chilled water and the other two-pipe system is for heating water. The HVACR unit typically has two separate water coils, one for heating water and one for chilled water. There is no mixing of the chilled and heated water. Each coil has its own automatic control valve, either two-way or three-way, and manual service valves. Some

HVACR units have only one coil. The coil is supplied with both heated and chilled water. A three-way valve is installed in the supply line to the coil. The chilled water and heated water supplies are not mixed. The three-way valve switches to deliver either chilled water or heating water (but not both) to the coil. In the return line from the coil a three-way, two-position valve diverts the leaving heated or chilled water to the correct return main.

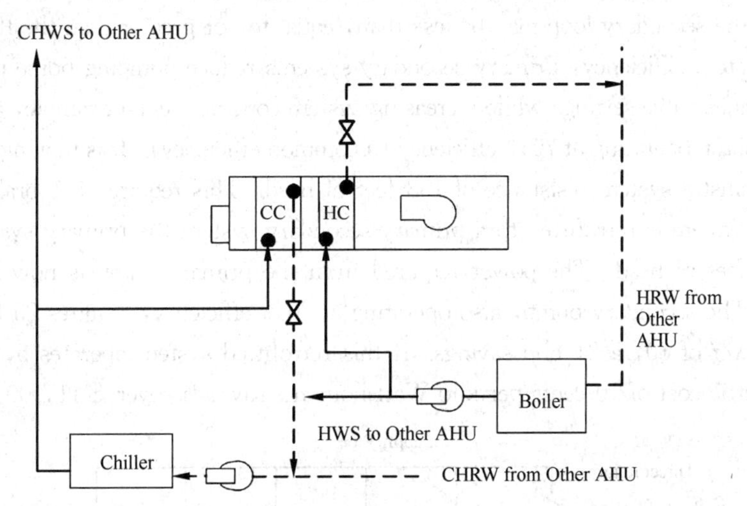

Figure 11-2 Four-pipe system with 2-way ATCV. AHU is a blow-thru system. CC: Cooling Coil, HC: Heating Coil, HWS: Heating Water Supply, HWR: Heating Water Return, CHWS: Chilled Water Supply, CHWR: Chilled Water Return.

Direct Return System and Reverse Return System

Both direct return and reverse return systems, as shown in Figure 11-3, have a supply pipe and a return pipe to each water coil and typically there will separate automatic temperature control valves and manual service and balancing valves for each water coil. For direct return systems the water coils are piped so that the first coil supplied by the boiler (or chiller) is also the first coil returned back to the boiler (or chiller). The reverse return systems have the water coils piped so that the first coil supplied is the last coil returned. The direct return system requires less main pipe than a reverse return systems and therefore has a lower initial piping cost. Reverse return systems require more piping than direct return systems and therefore have a higher initial piping cost but are considered to be self-balancing or require less balancing. System Efficiency: Flow meters and balancing valves are needed throughout to water balance either system, including bypasses.

Primary-Secondary System

In a primary-secondary system the primary pump circulates the water though the primary (main) loop and the secondary pump moves water through the secondary loop. A primary-secondary system can have one or many secondary circuits. In order to overcome the pressure loss in the secondary circuit and provide water flow to the coil a secondary pump is installed. When the two circuits are interconnected the primary pump and the secondary pump haveon effect on each other i. e., flow in one will not cause flow in the other. A crossover pipe, which may vary in length to a maximum of about two feet, is installed between the primary and secondary loop. The cross-over pipe has a negligible pressure drop which ensures the isolation of the two loops. The water flow in the secondary loop may be less than, equal to, or greater than the flow in the primary loop. System Efficiency: Primary-secondary systems reduce pumping horse power requirements and balance valve settings while increasing system control. As an example, a primary only system has a pump operating at 70% efficiency (a common efficiency). It is moving 1,825 gallons per minute against a system resistance of 153 feet of head. This requires 101 brake horsepower (bhp). If the system is retrofitted to a primary-secondary system the primary system resistance reduces to 83 feet of head. The power required from the primary pump is now only 55 brake horse power. The secondary pump, also operating at 70% efficiency, requires 25 bhp for a total brake horsepower of 80, a 21 bhp savings. If this retrofitted system operates 5,000 hours per year at an electric cost of 10 cents per kilo Watt-hour the saving is over $11,000 per year.

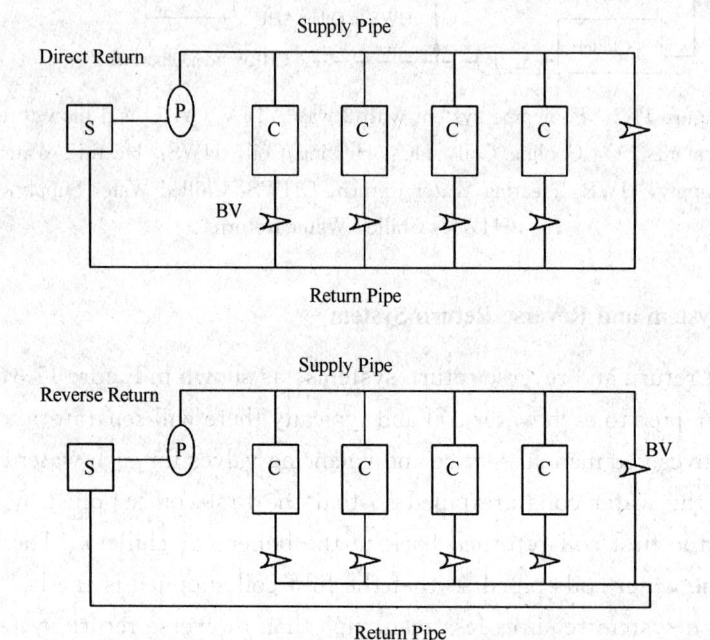

Figure 11-3 Direct return system and reverse return system. S-chiller or Boiler, P-Pump, C-Coil, BV-Manual Balancing Valve

11.2 Water Piping

Piping Material

For water systems, the piping materials most widely used are steel, both black (plain) and galvanized (zinc-coated), in the form of either welded-seam steel pipe or seamless steel pipe; ductile iron and cast iron; hard copper; and polyvinyl chloride (PVC). The piping materials for various services are shown inTable 11-1.

Copper, galvanized steel, galvanized ductile iron, and PVC pipes have better corrosion resistance than black steel pipes. Technical requirements, as well as local customs, determine the selection of piping materials.

The piping materials for various services. **Table 11-1**

Service	Piping Material
Chilled water	Black and galvanized steel
Hot water	Black steel, hard copper
Cooling water and drains	Black steel, galvanized ductile iron, PVC

Piping Dimensions

The steel pipe wall thicknesses currently used were standardized in 1930. The thickness ranges from Schedule 10, light wall, to Schedule 160, very heavy wall. Schedule 40 is the standard for a pipe with a diameter up to 10 inch. (250 mm). For instance, a 2-inch. (50-mm) standard pipe has an outside diameter of 2. 375 inch. (60. 33 mm) and an inside diameter of 2. 067 inch. (52. 50 mm). The nominal pipe size is only an approximate indication of pipe size, especially for pipes of small diameter. The dimensions of commonly used steel pipes are listed in ASHRAE Handbook.

The outside diameter of extruded copper is standardized so that the outside diameter of the copper tubing is 1/8 inch. (3. 2 mm) larger than the nominal size used for soldered or brazed socket joints. As in the case with steel pipes, the result is that the inside diameters of copper tubes seldom equal the nominal sizes. Types K, L, M, and DWV designate the wall thickness of copper tubes: type K is the heaviest, and DWV is the lightest. Type L is generally used as the standard for pressure copper tubing. Type DWV is used for drainage at atmospheric pressure.

Copper tubes are also categorized as hard and soft copper. Soft pipes should be used in applications for which the pipe will be bent in the field. The dimensions of copper tubes are given in ASHRAE Handbook. Thermoplastic plastic pipes are the most widely used plastic pipes in air conditioning. They are manufactured with dimensions that match steel pipe dimensions. The advantages of plastic pipes include resistance to corrosion, scaling, and the growth of algae and fungi. Plastic pipes have smooth surfaces and negligible age allowance. Age allowance is the allowance for corrosion and scaling for plastic pipes during their service life. Most plastic pipes are low in cost, especially compared with corrosion-resistant metal tubes.

The disadvantages of plastic pipes include the fact that their pressure ratings decrease rapidly when the water temperature rises above 100℉ (37. 8℃). PVC pipes are weaker than metal pipes and must usually be thicker than steel pipes if the same working pressure is to be maintained. Plastic pipes may experience expansion and contraction during temperature changes that is 4 times greater than that of steel. Precautions must be taken to protect plastic pipes from external damage and to account for its behavior during fire. Some local codes do not permit the use of some or all plastic pipes. It is necessary to check with local authorities.

Pipe Joints

Steel pipes of small diameter (2 inch. or 50 mm less) threaded through cast-iron fittings are the most widely used type of pipe joint. For steel pipes of diameter 2 inch. (50 mm) and more, welded joints, bolted flanges, and grooved ductile iron joined fittings are often used. Galvanized steel pipes are threaded together by galvanized cast iron or ductile iron fittings.

Copper pipes are usually joined by soldering and brazing socket end fittings. Plastic pipes are often joined by solvent welding, fusion welding, screw joints, or bolted flanges.

Vibrations from pumps, chillers, or cooling towers can be isolated or dampened by means of flexible pipe couplings. Arch connectors are usually constructed of nylon, dacron, or polyester and neoprene. They can accommodate deflections or dampen vibrations in all directions. Restraining rods and plates are required to prevent excessive stretching. A flexible metal hose connector includes a corrugated inner core with a braided cover. It is available with flanged or grooved end joints.

Working Pressure and Temperature

In a water system, the maximum allowable working pressure and temperature are not limited to the pipes only; joints or the pipe fittings, especially valves, may often be the weak links. The types of pipes, joint, and fittings and their maximum allowable working pressures for specified temperatures are given in relative handbook.

Expansion and Contraction

During temperature changes, all pipes expand and contract. The design of water pipes must take into consideration this expansion and contraction. Both the temperature change during the operating period and the possible temperature change between the operating and shutdown periods should also be considered. For chilled and condenser water, which has a possible temperature change of 40℉ to 100℉ (4.4℃ to 37.8℃), expansion and contraction cause considerable movement in a long run of piping. Unexpected expansion and contraction cause excess stress and possible failure of the pipe, pipe support, pipe joints, and fittings.

Expansion and contraction of hot and chilled water pipes can be better accommodated by using loops and bends. The commonly used bends are U bends, Z bends, and L bends, as shown in Figure 11-4. Anchors are the points where the pipe is fixed so that it will expand or contract between them. Reaction forces at these anchors should be considered when the support is being designed. *ASHRAE Handbook* 1992, *HVAC Systems and E-*

quipment, gives the required calculations and data for determining these stresses. Guides are used so that the pipes expand laterally.

Empirical formulas are often used instead of detailed stress analyses to determine the dimension of the offset leg L_0[ft (m)]. Waller (1990) recommended the following formulas:

$$\begin{aligned} \text{U Bends:} L_o &= 0.041 D^{0.48} L_{ac}^{0.46} \Delta T \\ \text{Z Bends:} L_o &= (0.13 D L_{ac} \Delta T)^{0.5} \\ \text{L Bends:} L_o &= (0.314 D L_{ac} \Delta T)^{0.5} \end{aligned} \tag{11-1}$$

where D —diameter of pipe, inch. (mm)

L_{ac} —distance between anchors, hundreds of ft (m)

ΔT —temperature difference, °F (℃)

If there is no room to accommodate U, Z, or L bends (such as in high-rise buildings or tunnels), mechanical expansion joints are used to compensate for movement during expansion. Packed expansion joints allow the pipe to slide to accommodate movement during expansion. Various types of packing are used to seal the sliding surfaces in order to prevent leakage. Another type of mechanical joint uses bellows or flexible metal hose to accommodate movement. These types of joints should be carefully installed to avoid distortion.

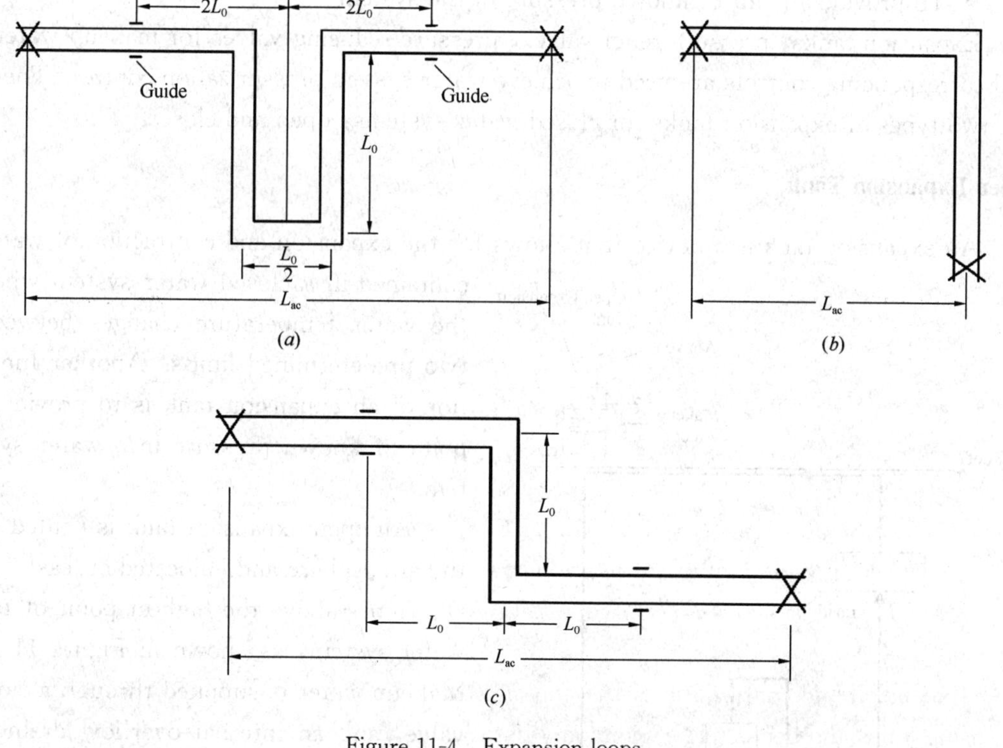

Figure 11-4 Expansion loops.
(*a*) U bends; (*b*) L bends; (*c*) Z bends

11.3 Pressurization and the Presence of Air

Water System Pressurization Control

For an open water system, the maximum operating gauge pressure is the pressure at a specific point in the system where the positive pressure exerted by the water pumps, to overcome the pressure drops across the equipment, components, fittings, and pipes plus the static head due to the vertical distance between the highest water level and that point, is at a maximum.

In a closed chilled or hot water system, a variation in the water temperature will cause an expansion of water that may raise the water pressure above the maximum allowable pressure. The purposes of system pressurization control for a closed water system are as follows:

- To limit the pressure of the water system to below its allowable working pressure
- To maintain a pressure higher than the minimum water pressure required to vent air
- To assist in providing a pressure higher than the net positive suction head (NPSH) at the pump suction to prevent cavitation
- To provide a point of known pressure in the system

Expansion tanks, pressure relief valves, pressure-reducing valves for makeup water, and corresponding controls are used to achieve water system pressurization control. There are two types of expansion tanks for closed water systems: open and closed.

Open Expansion Tank

An expansion tank is a device that allows for the expansion and contraction of water contained in a closed water system when the water temperature changes between two predetermined limits. Another function of an expansion tank is to provide a point of known pressure in a water system.

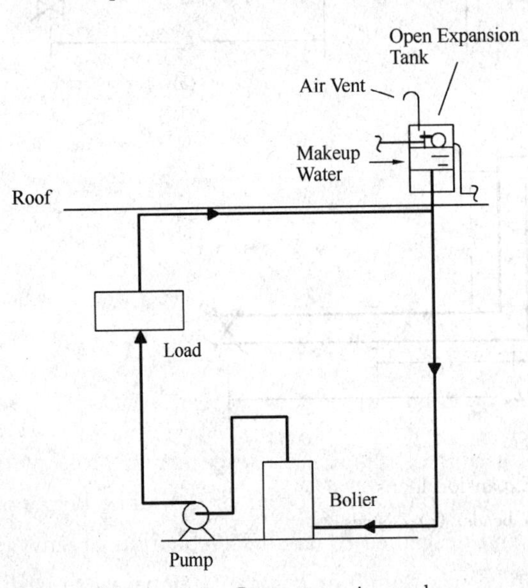

Figure 11-5 Open expansion tank.

An open expansion tank is vented to the atmosphere and is located at least 3 ft (0.91 m) above the highest point of the water system, as shown in Figure 11-5. Makeup water is supplied through a float valve, and an internal overflow drain is always installed. A float valve is a globe or ball valve connected with a float ball to regulate the makeup water flow according to the liquid level in the tank. An open expansion tank is often connected to the suction side of the water pump to prevent the water pressure in the system from dropping below the atmospheric pressure. The pressure of the liquid level in the open tank is equal to the atmospheric pressure, which thus provides a reference point of known pressure to determine the water pressure at any point in the water system. The minimum tank volume should be at least 6 percent of the volume of water in the system V_s, ft^3 (m^3). An open expansion tank is simple, more stable in terms of system pressure characteristics, and low in cost. If it is installed indoors, it often needs a high ceiling. If it is installed outdoors, water must be prevented from freezing in the tank, air vent, or pipes connected to the tank when the outdoor temperature is below 32°F (0℃). Because the water surface in the tank is exposed to the atmosphere, oxygen is more easily absorbed into the water, which makes the tank less resistant to corrosion than a diaphragm tank (to be described later). Because of these disadvantages, an open expansion tank has only limited applications.

Closed Expansion Tank

A closed expansion tank is an airtight tank filled with air or other gases, as shown in Figure 11-6. When the temperature of the water increases, the water volume expands. Excess water then enters the tank. The air in the tank is compressed, which raises the system pressure. When the water temperature drops, the water volume contracts, resulting in a reduction of the system pressure.

To reduce the amount of air dissolved in the water so as to prevent corrosion and prevent air noise, a diaphragm, or a bladder, is often installed in the closed expansion tank to separate the filled air and the water permanently. Such an expansion tank is called a diaphragm, or bladder, expansion tank. Thus, a closed expansion tank is either a plain closed expansion tank, which does not have a diaphragm to separate air and water, or a diaphragm tank.

For a water system with only one air-filled space, the junction between the closed expansion tank and the water system is a point of fixed pressure. At this point, water pressure remains constant whether or not the pump is operating because the filled air pressure depends on only the volume of water in the system. The pressure at this point can be determined according to the ideal gas law, as given by Equation (4-3): $pv = RT_R$. The pressure in a closed expansion tank during the initial filling process or at the minimum operating pressure is called the fill pressure p_{fil}, psia. The fill pressure is often used as the reference pressure to determine the pressure characteristics of a water system.

Air in Water Systems

In a closed recirculated water system, air and nitrogen are present in the following forms: dissolved in water, free air or gas bubbles, or pockets of air or gas. The behavior of air or gas dissolved in liquids is governed and described by Henry's equation. Henry's equation states that the amount of gas dissolved in a liquid at constant temperature is directly proportional to the absolute pressure of that gas acting on the liquid, or

$$x = \frac{p}{H} \tag{11-2}$$

where x——amount of dissolved gas in solution, percent by volume

p——partial pressure of that gas, psia

H——Henry's constant; changes with temperature

The lower the water temperature and the higher the total pressure of the water and dissolved gas, the greater the maximum amount of dissolved gas at that pressure and temperature.

When the dissolved air or gas in water reaches its maximum amount at that pressure and temperature, the water becomes saturated. Any excess air or gas, as well as the coexisting water vapor, can exist only in the form of free bubbles or air pockets. A water velocity greater than 1.5 ft/s (0.45 m/s) can carry air bubbles along with water. When water is in contact with air at an air-water interface, such as the filled airspace in a plain closed expansion tank, the concentration gradient causes air to diffuse into the water until the water is saturated at that pressure and temperature. An equilibrium forms between air and water within a certain time. At specific conditions, 24 h may be required to reach equilibrium.

The oxygen in air that is dissolved in water is unstable. It reacts with steel pipes to form oxides and corrosion. Therefore, after air has been dissolved in water for a long enough time, only nitrogen remains as a dissolved gas circulating with the water.

Penalties due to Presence of Air and Gas

The presence of air and gas in a water system causes the following penalties for a closed water system with a plain closed expansion tank:

- Presence of air in the terminal and heat exchanger, which reduces the heat-transfer surface
- Corrosion due to the oxygen reacting with the pipes
- Waterlogging in plain closed expansion tanks
- Unstable system pressure
- Poor pump performance due to gas bubbles
- Noise problems

There are two sources of air and gas in a water system. One is the air-water interface in a plain closed expansion tank or in an open expansion tank, and the other is the dissolved air in a city water supply.

Oxidation and Waterlogging

Consider a chilled water system that uses a plain closed expansion tank without a diaphragm, as shown in Figure 11-6. This expansion tank is located in a basement, with a water pressure of 90 psig (620 kPa · g) and a temperature of 60℉ (15.6℃) at point A.

At such a temperature and pressure, the solubility of air in water is about 14. 2 percent. The chilled water flows through the water pump, the chiller, and the riser and is supplied to the upper-level terminals. During this transport process, part of the oxygen dissolved in the water reacts with the steel pipes to form oxides and corrosion. At upper-level point B, the water pressure is only 10 psig (69 kPa • g) at a chilled water temperature of about 60°F (15. 6℃). At this point, the solubility of air in water is only about 3. 3 percent. The difference in solubility between point A and B is 14. 2−3. 3=10. 9 percent. This portion of air, containing a higher percentage of nitrogen because of the formation of oxides, is no longer dissolved in the chilled water, but is released from the water and forms free air, gas bubbles, or pockets. Some of the air pockets are vented through air vents at the terminals, or high points of the water system. The chilled water returns to point A again and absorbs air from the air-water interface in the plain closed expansion tank, creating an air solubility in water of about 14. 2 percent. Of course, the actual process is more complicated because of the formation of oxides and the presence of water vapor.

Such a chilled waterrecirculating process causes the following problems:

- Oxidation occurs because of the reaction between dissolved oxygen and steel pipes, causing corrosion during the chilled water transport andrecirculating process.
- The air pockets vented at high levels originally come from the filled air in the plain closed expansion tank; after a period of recirculation of the chilled water, part of the air charge is removed to the upper levels and vented. The tank finally waterlogs and must be charged with compressed air again. Waterlogging also results in an unstable system pressure because the amount of filled air in the plain closed expansion tank does not remain constant. Oxidation and water logging also exist in hot water systems, but the problems are not as pronounced as in a chilled water system.

Oxidation and waterlogging can be prevented or reduced by installing a diaphragm expansion tank instead of a plain closed expansion tank. Air vents, either manual or automatic, should be installed at the highest point of the water system and on coils and terminals at higher levels if a water velocity of not less than 2 ft/s (0. 61 m/ s) is maintained in the pipes, in order to transport the entrained air bubbles to these air vents.

In a closed chilled water system using a diaphragm expansion tank, there is no air-water interface in the tank. The 3. 3 percent of dissolved air, or about 2. 6 percent of dissolved nitrogen, in water returning from point B to A cannot absorb more air again from the diaphragm tank. If there is no fresh city water supply to the water system, then after a period of water recirculation the only dissolved air in water will be the 2. 6 percent nitrogen. No further oxidation occurs after the initial dissolved oxygen has reacted with the steel pipe. Waterlogging does not occur either.

Because of the above concerns, a closed water system should have a diaphragm or bladder expansion tank. An open expansion tank at high levels causes fewer problems than a plain closed expansion tank. A diaphragm tank may be smaller than an equivalent plain tank. An air eliminator or air separator is usually required for large water systems using a

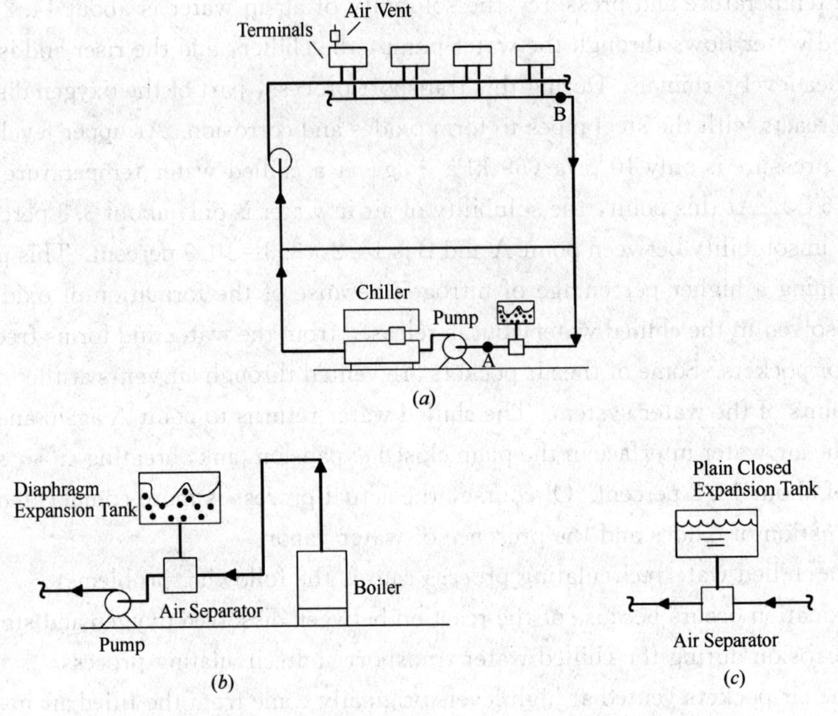

Figure 11-6 Closed expansion tank for a water system

(a) Diaphragm expansion tank in a chilled water system; (b) Diaphragm expansion tank in a hot water system; (c) Plain closed expansion tank

diaphragm tank to separate dissolved air from water when the water system is charged with a considerable amount of city water.

Unit 12 Fans

A fan is the prime mover of an air system or ventilation system. It moves the air and provides continuous airflow so that the conditioned air, space air, exhaust air, or outdoor air can be transported from one location to another through air ducts or other air passages.

A fan is also a turbomachine in which air is usually compressed at a compression ratio R_{com} not greater than 1.07. The compression ratio, dimensionless, is defined as

$$R_{com} = \frac{p_{dis}}{p_{suc}} \tag{12-1}$$

where p_{dis}——discharge pressure at outlet of compressor or fan, lb_f/in^2. abs. or psia (kPa abs.)

p_{suc}——suction pressure at inlet of compressor or fan, psia (kPa abs.)

A blower is usually an enclosed multiblade rotor that compresses air to a higher discharge pressure. There is no clear distinction between a fan and a blower. Traditionally, blowers do not discharge air at low pressure as some fans do. A fan is driven by a motor directly (direct drive) or via belt and pulleys (belt drive). Some large industrial fans in power plants are driven by steam or gas turbines.

12.1 Types of Fan

There are two distinctly different types of fan design based on the direction of airflow through the fan: centrifugal and axial. Air flows into a centrifugal fan parallel to the motor shaft and is then spun outward, through the fan blades, by the centrifugal motion of the fan. The fan housing then channels the airflow toward the fan outlet. The air movement in an axial fan is parallel to the shaft. A propeller fan is an example of an axial fan. Propeller fans are often used in residential condensing units.

Centrifugal fans are designed to produce high static pressures, while axial fans provide high volumes of air. Centrifugal fans work very well on duct air-distribution systems, while in contrast axial fans are not capable of producing enough static pressure to move air through long ducts. However, axial fans are very effective in moving air through the large openings typically required for building ventilation. Each fan type has its purpose and application, and they are not interchangeable.

Centrifugal Fans

Centrifugal fans are the workhorse of the air-conditioning industry. These fans are sometimes called "squirrel cage" fans due to the shape of the impeller wheel. They are the fan of choice when "pumping" air in high volumes, against high static pressure. A ductless system may use a centrifugal fan to deliver air directly to a room or use a blower in a split system to pump air through an elaborate duct system to condition an entire residence. These fans are also used in many ventilation applications.

There are several different blade designs used with centrifugal fans. The two main types of centrifugal fans used in air-conditioning work are the forward-curved blade, which is most common, and the backward-curved blade.

The advantage of the backward-curved blade is that it is nonoverloading. The disadvantage is that it is noisier.

Centrifugal Fan Housings

Centrifugal fans can be divided into three housing types: scroll, tubular, and plug, the last of which has no housing. The scroll housing is the most widely used for air conditioning and heating. This housing design directs the airflow from the fan blades down a channel at a 90-degree angle from the drive shaft. Figure 12-1 shows a centrifugal fan with a scroll housing.

Figure 12-1 Centrifugal fan with a scroll housing.

Tubular centrifugal fans have a housing that looks like a tube or pipe. The air is drawn in parallel to the drive shaft and spun out of the blades. It is then redirected out the other end of the "tube," giving it parallel airflow to the drive shaft when exiting the fan housing. This centrifugal fan is used to ventilate heat, fumes, or smoke in commercial or industrial applications.

The third and least used centrifugal fan is called a plug fan. It is a design that has no housing. These fans are installed directly into plenums of AHUs to pressurize the complete unit. This fan design is less efficient but offers flexibility in unit design and can reduce unit size.

The amp draw of a centrifugal blower will actually decrease if the air to the blower is blocked. A centrifugal fan will turn faster if it is not actually moving air, and the motor amp draw decreases because of the reduced load on the blower.

Forward-Curved Centrifugal Fans

Forward-curved centrifugal fans are the most popular blade design for residential systems. This blade design curves the blade's leading edge toward the direction of airflow. This adds resistance to the forward motion of the fan wheel, reducing fan speed, noise, and efficiency. This blade design can cause motor overloading if the static pressure is decreased past design limits of the fan. One advantage to this design is that bearing size is reduced and bearing life extended because of the fan's lower rpm.

The indoor blower in most furnaces and heat pumps is a forward-curved centrifugal fan. These blowers are not designed to operate in free air; they need to have some resistance to airflow. Technicians often try and use blowers removed from old equipment as fans to cool the work area. Some restriction must be added to the opening of either the fan inlet or outlet to keep the fan operating without overloading the motor.

Backward-Curved Centrifugal Fans

Backward-curved fan blades have a leading edge that trails the airflow, producing little resistance to the fan wheel's forward motion. The speed of one of these blowers is nearly twice the speed of a forward-curved blower. This lowresistance design increases both efficiency and noise due to its higher rpm. These fans are larger in design and require larger drive shafts and bearings than do forward-curved blowers. A comparison of forward- and backward-curved blades is shown in Figure 12-2.

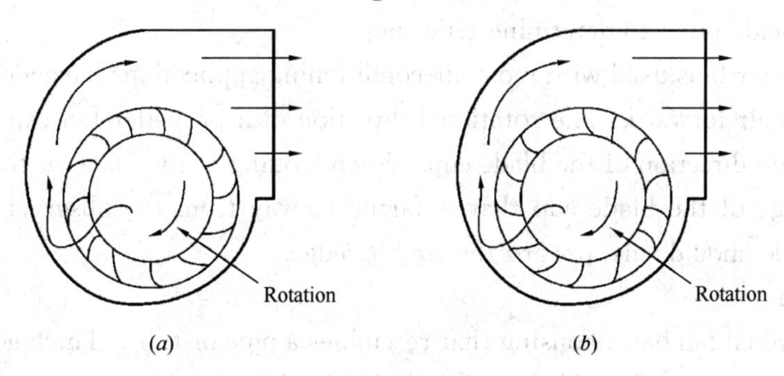

Figure 12-2 Schematic view of the construction of fan
(*a*) backward-curved fan; (*b*) forward-curved fan

Airfoil Centrifugal Fans

Airfoil blade blowers are a variant of the backward-blade design. This is the most efficient centrifugal blade design due to its very low resistance to airflow. However, this resulting higher efficiency covers only a small area of its performance curve in relation to other backward-curved fan blades; and has a higher cost associated with it. This often reduces its use to only high-static-pressure applications where additional cost can be justified. In most other ways, the airfoil blade design reflects the backward-blade design's operational advantages.

Axial (in Line) Fans

Axial fans are divided into three types based on housing design: the propeller fan, the tube axial, and the vane axial. The propeller type will handle large volumes of air for low-pressure applications. It has high usage for exhaust fans and condenser fans. Vane-type fans are highly efficient but noisy. The blade pitch can be adjusted to control the amount of air the fan handles.

The amp draw of a propeller fan will increase if the air leaving the blower is blocked. Propeller fans do not work well against the high-pressure difference created when the fan outlet is blocked. They slow down, putting moreload on the fan motor and increasing its amp draw.

Propeller Fans

The most familiar axial fan consists of a fan shaft with two or more blades attached at a 90-degree angle to the shaft. Some of these fans will have a hub assembly that attaches the blades to the shaft. This hub assembly allows for maintenance and modification. Air is pulled through the fan blades parallel to the shaft and is blown outward. These fans are high speed and as a result produce high noise levels. Axial fans operate well in low-static environments and can move high volumes of air.

The propeller fan is generally is used as a condenser fan on air-conditioning systems. These fans are generally shrouded to improve performance but do not use a true housing. Propeller fans typically have two to six blades. This fan works well in a low static pressure and relies on blade pitch to determine efficiency.

The propeller fans used with most air-conditioning applications have a cup shape. The cup pushes the air forward. The rotational direction of a propeller fan can be determined by observing the direction of the blade cup. When looking at the blade so that the cups are visible, the edge of the blade cup that is farthest away from the observer is the leading edge. The blade should turn toward the leading edge.

Tube Axial Fan

The tube axial fan has a housing that resembles a pipe or tube. The blades are in close tolerance to the housing. Flat blade or foil design blades are most common. These fans are widely used in ventilation, such as warehouses or in structures where fume venting is required.

Vane Axial Fan

The vane axial fan is the most efficient axial design and can operate at higher static pressures than other axial designs. The efficiency of this housing comes from vanes added in the path of the airflow to reduce "swirling" of the air. These fans move high volumes of air and produce high noise levels and may require noise dampers in some applications. This type of design can offer an adjustable blade pitch that can be used to vary the amount of air moved by the fan. These fans are often applied to variable-air-volume systems because of their ability to easily vary the airflow.

The Air Movement and Control Association (AMCA) is an international standards organization that certifies fans, dampers, and other air-movement devices. The AMCA has classified fans into four classes based on fan performance. These classifications define upper and lower ranges of static pressure and outlet velocity for each class. Class 1 fans operate at the lowest pressures and velocities; Class 4 fans operate at the highest pressures and velocities. The AMCA standards require that the fan operate safely on or below the minimum rating for its class.

Special Design Fans

Centrifugal power roof ventilator and axial power roof ventilator are small fans used to exhaust air from restrooms, attic spaces, and other small areas. They operate at low static pressure, horsepower, and efficiency.

For tubular centrifugal fan (TCF) is typically used in return systems where saving space is a consideration.

Wheel Construction: The wheel is housed in a cylindrical tube with backward inclined or airfoil blades. Basically it is a hybrid in that it is a centrifugal fan wheel in a vane axial fan housing.

Static Pressure Range: Low.

Airflow and Discharge: The air is discharged radially from the wheel then changes directions by 90 degrees to flow through a guide vane section and then flows parallel to the fan shaft.

Efficiency: Lower than backward curved or backward inclined fan.

Horsepower Characteristics: The horse power curve increases with an increase in air quantity (but only to a point) to the right of maximum efficiency and then gradually decreased. "Non-overloading."

Performance Curve Characteristics: Generally, no dip in the pressure curve left of the peak pressure point, although some may have a dip similar to the axial fan. Stable and predictable operation is similar to the backward bladed fan except for lower capacities, pressures and efficiencies.

12.2 Fan Parameters

Fan Capacity or Volume Flow Rate

Fan capacity or the fan volume flow rate $\dot{V}_f$, in cfm (m^3/s), is defined as the rate of volume flow measured at the inlet of the fan, corresponding to a specific fan total pressure. It is usually determined by the product of the duct velocity and the area of the duct connected to the fan inlet, according to the test standard AMCA Standard 210-85 and ASHRAE Standard 51-1985.

Fan volume flow rate is independent of air density ρ. However, fan total pressure is affected by air density. Therefore, the fan volume flow rate is normally rated at standard air conditions, i. e., dry air at an atmospheric pressure of 14.696 psia (101,325 Pa abs.), a temperature of 70°F (21.1℃), and a density of 0.075 lb/ft³ (1.2 kg/m³).

Fan Pressure

Fan total pressure Δp_{tf}, expressed in inches of height of water column (inches WC or Pa), is the total pressure rise of a fan, i. e., the pressure difference between the total pressure at the fan outlet pto and the total pressure at the fan inlet pti, both in in. WG (Pag), or

$$\Delta p_{tf} = p_{to} - p_{ti} \tag{12-2}$$

Fan velocity pressure p_{vf}, in in. WC (Pa), is the pressure calculated according to the mean velocity at the fan outlet v_o, in fpm (m/s). If air density ρ= 0.075 lb/ft³ (1.2 kg/m³), it can be calculated as

$$p_{vf} = p_{vo} = \frac{\rho v_o^2}{2g_c} = \left(\frac{v_o}{4005}\right)^2 = \left(\frac{\dot{V}_o}{4005A_o}\right)^2 \tag{12-3}$$

where p_{vo}——velocity pressure at fan outlet, in. WC (Pa)

g_c——dimensional constant, 32.2 lbm · ft/lbf · s² (kg · m/N · s²)

$\dot{V}_o$——volume flow rate at fan outlet, cfm (m³/s)

A_o——cross-sectional area of fan outlet, ft² (m²)

Fan static pressure Δp_{sf}, in in. WC (Pa), is the difference between the fan total pressure and fan velocity pressure, or

$$\Delta p_{sf} = \Delta p_{tf} - p_{vf} = p_{to} - p_{ti} - p_{vo} = p_{so} - p_{to} \tag{12-4}$$

Fan Power and Fan Efficiency

Air power Pair, in hp (W), is the work done in moving the air along a conduit against a fan total pressure Δp_{tf}, in in. WC (Pa), at a fan volume flow rate of $\dot{V}_f$ in cfm (m³/s). Because 1 hp = 33,000 ft · lbf/min and 1 in. WC = 5.192 lbf/ft²,

$$p_{air} = \frac{\Delta p_{tf} \times 5.19 \times \dot{V}_f}{33,000} = \frac{\Delta p_{tf}\dot{V}_f}{6356} \tag{12-5}$$

The fan power input on the fan shaft, often called the brake horsepower P_f, can be calculated as

$$P_f = \frac{\Delta p_{tf}\dot{V}_f}{c\eta_t} = \frac{\Delta p_{tf}\dot{V}_f}{6353\eta_t} \tag{12-6}$$

When Δp_{tf} is expressed in Pa, fan volume flow rate in m³/s, fan total efficiency η_t is the ratio of air power Pair to fan power input P_f on the fan shaft, dimensionless, then constant C = 1, and fan power input P_f is expressed in W. If Δp_{tf} is expressed in in. WC, fan volume flow rate is in cfm, fan total efficiency η_t is a dimensionless ratio, then constant C= 6356 and fan power input on the shaft P is in hp.

From Equation (12-6), fan total efficiency can also be calculated as

$$\eta_t = \frac{\Delta p_{tf}\dot{V}_f}{6353P_f} \tag{12-7}$$

Fan total efficiency is a combined index of aerodynamic, volumetric, and mechanical efficiencies of a fan. Fan static efficiency η_s is defined as the ratio of the product of the fan static pressure Δp_{sf}, in in. WC, and the fan volume flow rate to the fan power input, i.e.,

$$\eta_s = \frac{\Delta p_{sf}\dot{V}_f}{6353P_f} \tag{12-8}$$

Air Temperature Increase through Fan

If air density ρ_a= 0.075 lb/ft³, the specific heat of air c_{pa}=0.243 Btu/lb · ℉, and

1 hp = 42.41 Btu/min, the relationship between fan power input and the air temperature increase when it flows through the fanΔT_{f}, in °F, is given as

$$P_{\mathrm{f}} = \frac{V_{\mathrm{f}}\rho_{\mathrm{a}}c_{\mathrm{pa}}\Delta T_{\mathrm{f}}}{42.41} \tag{12-9}$$

Combining Equation (12-6) and Equation (12-9), then, gives

$$\Delta T_{\mathrm{f}} = \frac{0.00667\Delta p_{\mathrm{tf}}}{\rho_{\mathrm{a}}c_{\mathrm{pa}}\eta_{\mathrm{t}}} = \frac{0.37\Delta p_{\mathrm{tf}}}{\eta_{\mathrm{t}}} \tag{12-10}$$

This air temperature rise in a fan when air flows through it is caused by the compression process and energy losses that occur inside the fan. When air flows through the air duct, duct fittings, and equipment, the duct friction loss and dynamic losses cause a temperature increase as mechanical energy is converted to heat energy. However, this temperature increase in the air duct is offset by a temperature drop caused by the expansion of air due to the reduction of static pressure along the airflow. Therefore, it is more convenient to assume that the air temperature rise occurs because of the friction and dynamic losses along the airflow only when air is flowing through the fan.

12.3 Fan Laws

A fan's performance varies depending upon the conditions it operates under. There is no such thing as a 400 CFM fan. A fan that moves 400 CFM of air at one condition might only move 200 CFM at another condition. The amount of aira fan moves, the speed of the fan, the amount of pressure difference produced across the fan, and the motor horsepower required are all interrelated.

Fan speed is directly related to the amount of air a fan can move: doubling the fan speed doubles the airflow. Take, for example, a fan moving 400 CFM against a static pressure of 0.1 in wc turning 500 rpm and using a¼-hp motor. Doubling the speed from 500 rpm to 1,000 rpm will double the CFM from 400 CFM to 800 CFM. The static pressure difference across the fan increases more rapidly. Doubling the fan speed increases the static pressure by a factor of 4. Doubling the speed of the same fan from 500 rpm to 1,000 rpm will increase the static pressure across the fan from 0.1 in wc to 0.4 in wc. The horsepower requirement to accomplish this increases even more dramatically. Doubling the fan rpm will require 8 times the horsepower! So increasing the fan speed from 500 rpm to 1,000 rpm will increase the horsepower requirement from ¼ hp to 2 hp.

The relationship between changes in speed (rpm), air volume (CFM), static pressure (SP), and power in brake horsepower (Bhp) are described by a set of formulas called fan laws. The fan laws in classic form are:

$$\frac{\mathrm{CFM_2}}{\mathrm{CFM_1}} = \frac{\mathrm{rpm_1}}{\mathrm{rpm_2}}$$

$$\frac{\mathrm{SP_2}}{\mathrm{SP_1}} = \left(\frac{\mathrm{rpm_1}}{\mathrm{rpm_2}}\right)^2 \tag{12-11}$$

$$\frac{Bhp_2}{Bhp_1} = \left(\frac{rpm_1}{rpm_2}\right)^3$$

These can be algebraically rearranged to find the new airflow, speed, static pressure, or horsepower.

$$\text{New CFM} = \text{Old CFM} \times \frac{\text{New RPM}}{\text{Old RPM}}$$

$$\text{New RPM} = \text{Old RPM} \times \frac{\text{New CFM}}{\text{Old CFM}}$$

$$\text{New SP} = \text{Old SP} \times \left(\frac{\text{New RPM}}{\text{Old RPM}}\right)^2 \qquad (12\text{-}12)$$

$$\text{New Bhp} = \text{Old Bhp} \times \left(\frac{\text{New RPM}}{\text{Old RPM}}\right)^3$$

EXAMPLE 12-1 Result from Increasing Fan CFM

What would be the resulting operating condition if the airfl ow moved by a fan operating at 300 rpm, 0. 15 in wc static, and 0. 2 Bhp was increased from 600 CFM to 800 CFM?

SOLUTION First find the new speed.

New rpm = 300 rpm × 800 CFM/600 CFM = 400 rpm

Next, use the new rpm to fi nd the new static pressure.

New sp = 0. 15 in wc × (400 rpm/300 rpm)^2 = 0. 15 in wc × 1. 78 = 0. 27 in wc static

Finally, use the new rpm to fi nd the new horsepower.

New Bhp = 0. 2 Bhp × (400 rpm/300 rpm)^3 = 0. 2 Bhp × 2. 37 = 0. 47 Bhp

The new operating conditions are at 400 rpm, 0. 27 in wc static, and 0. 47 Bhp to increase the airfl ow from 600 CFM to 800 CFM.

Fan Performance Tables

Manufacturers publish performance tables for each specific fan. Performance tables show the fan performance at specific operating conditions. Performance tables show the fan CFM compared to the static pressure. Other information included in fan performance tables are the outlet velocity, fan rpm, and motor horsepower requirement for the selected condition.

You can use the tables to help determine how the fan is operating under field conditions by measuring the fan speed and the fan static pressure and entering this information on the table. If the measured conditions are within the scope of the table the approximate CFM and brake horsepower can be determined.

Fans that have high rotating speeds and operate at high pressures are built to withstand the stresses of centrifugal force. However, if the fan is rotating too fast, the wheel could fly apart or the fan shaft could whip. Therefore, for safety reasons, the performance tables generally list the maximum RPM for each class of fan. Maximum allowable fan speed should be checked before increasing the fan RPM to ensure that the new operating condition doesn't require a different class fan.

Very often the fans, motors, and drives are all located in the airstream. If this is the case, the heat from the motor, the frictional heat from the belt drives, and the heat from the bearings need to be added to the total cooling load capacity.

Fan Performance Curves

The fan curve is a graphical representation of fan performance. Airflow is typically plotted along the x-axis, and power, pressure, and efficiency are plotted along the y-axis as shown in Figure 12-3. Fan curves are used in system design to select fan that is capable of moving the correct amount of air at the desired system operating condition. There will be a best match for air flow, efficiency, and power for any given fan. Fan curves can also be used to predict the effect any change in operating conditions will have on fan performance. A fan or a family of fans may be compared on one chart.

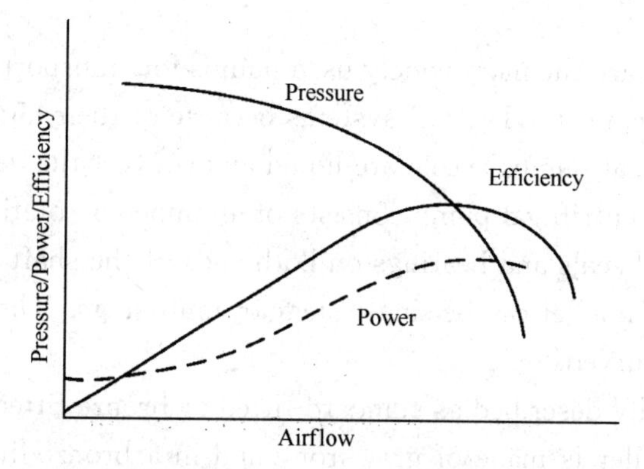

Figure 12-3 Typical fan curve with CFM along the bottom and all other data on the side.

Unit 13 Pumps

The prime mover of water in HVACR systems is motor driven water pumps. Water and air are fluids, air is compressible and water is not. The physical laws of water and air are the same and the prime movers and other components are essentially the same, the names and terms, however, change (fan wheel, pump impeller, air duct, water pipe, air damper, water valve, cubic feet per minute for air volume, gallons per minute for water volume etc.).

13.1 Centrifugal Pumps

Centrifugal pumps are the most widely used pumps for transporting chilled water, hot water, and condenser water in HVACR systems because of their high efficiency and reliable operation. Centrifugal pumps accelerate liquid and convert the velocity of the liquid to static head. A typical centrifugal pump consists of an impeller rotating inside a spiral casing, a shaft, mechanical seals and bearings on both ends of the shaft, suction inlets, and a discharge outlet. The impeller can be single-stage or multistage. The vanes of the impeller are usually backward curved.

The pump is usually described as standard-fitted or bronze-fitted. In a standard-fitted construction, the impeller is made of gray iron, and in a bronze-fitted construction, the impeller is made of bronze. For both constructions, the shaft is made of stainless steel or alloy steel, and the casing is made of cast iron.

Three types of centrifugal pumps are often used in water systems in HVACR systems: double-suction horizontal split-case, frame-mounted end suction, and vertical in-line pumps. Double-suction horizontal split-case centrifugal pumps are the most widely used pumps in large central hydronic air conditioning systems.

These are referred to as non-positive displacement pumps. Unlike a gear-type pump, a centrifugal pump can be operated with the discharge valve closed. This is called the pump shutoff head. However, if the centrifugal pump operates for an extended period of time with the discharge valve closed, it will overheat due to the fluid friction developed.

Volume Flow Rate

Volume flow rate $\dot{V}_p$ is the capacity handled by a centrifugal pump. The pump motor spins an impeller, which can be similar to a flat disk with vanes. This is called an open impeller. There are many different sizes and types of impellers. The spinning impeller increases the water velocity, thereby making it flow out of the pump and through the piping.

The flow of water is measured in gallons per minute (gpm).

Head Pressure

The weight of water in a pipe or tank exerts a force. As the height of water in a pipe or tank becomes greater, the force increases. This force can be measured as a pressure and is often referred to as head pressure. The units of head pressure are commonly expressed in feet rather than psi. The height of the water column is used to determine the head pressure. One foot of water exerts a pressure equivalent to 0.433 psi. The pressure as measured by a gauge at the bottom of an open water tank 20 ft high would be equal to:

$$20\ \text{ft.} \times 0.433\ \text{psi/ft} = 8.66\ \text{psi} \tag{13-1}$$

The solution shows that a pressure gauge at the bottom of an open water tank 20 ft high would read 8.66 psig. The diameter of the tank does not matter, just the height as shown in Figure 13-1. This is because the pressure measured is per square inch, and 1 in 2 of area in a 10-ft diameter tank is the same as 1 in 2 area of a 5-ft diameter tank. It must be understood then, that the head pressure of a 1-in diameter pipe is the same as the head pressure in a ½-in-diameter pipe if both are 20 ft high. For quick estimates, the relationship between water height and pressure is almost two to one. 100 feet of water would have a pressure at the bottom of about 50 psig. However realize that the more accurate measurement would be 100 × 0.433 = 43.3 psig.

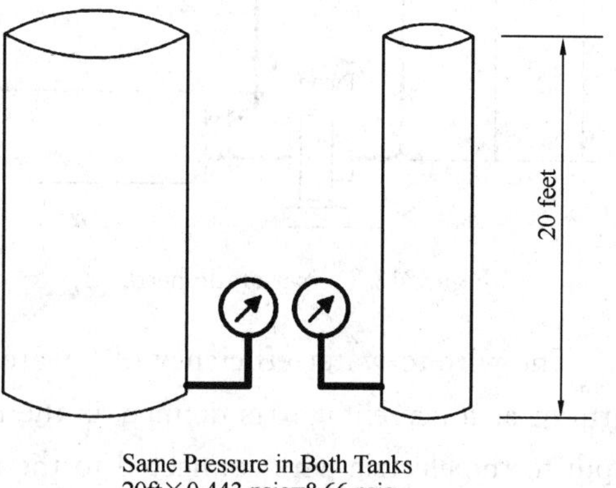

Figure 13-1 Relationship between tank size and head pressure.

On a cross-sectional plane perpendicular to fluid flow in a water system, the static head H_s[ft (m)] is the pressure expressed in feet (meters) of water column that is exerted on the surrounding fluid and enclosure. On a cross-sectional plane, velocity head H_v[ft (m)] can be calculated as

$$H_v = \frac{v_o^2}{2g} \tag{13-2}$$

where v_o——velocity of water flow at pump outlet, ft/s (m/s)

g——gravitational acceleration, 32.2 ft/s^2 (9.81 m/s^2)

Total head H_t[ft (m)] is the sum of static head and velocity head, i. e.,

$$H_t = H_s + H_v \tag{13-3}$$

Net static head ΔH_s[ft (m)] is the head difference between the discharge static head H_{dis} and suction static head H_{suc}, both in feet (meter), as shown in Figure 13-2.

Power and Efficiency

Pump power P_p(hp) is the power input on the pump shaft; and pump efficiency η_p is

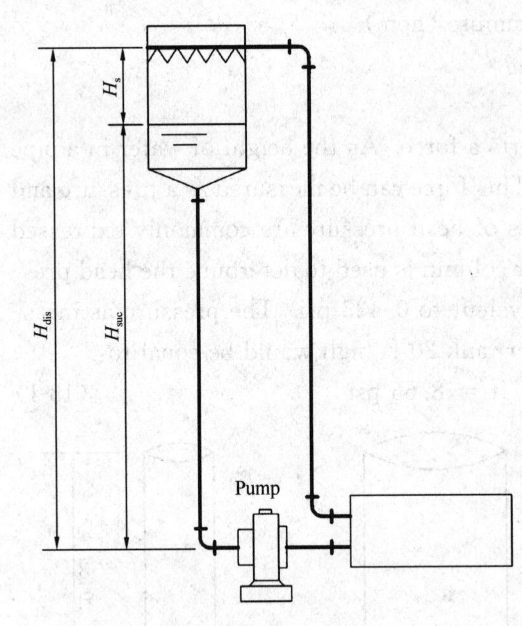

Figure 13-2 Net static head.

the ratio of the energy output from water to the power input on the pump shaft, and both can be calculated as

$$P_p = \frac{V_p H_t g_s}{3960\eta_p}$$
$$\eta_p = \frac{V_p H_t g_s}{3960 P_p} \tag{13-4}$$

where g_s = specific gravity, i. e., the ratio of the mass of liquid handled by the pump to the mass of water at 39℉(4℃).

A pump may be directly driven by a motor, or it may be driven by a motor and belts. When the energy cost of a water system is evaluated, the pump total efficiency η_p, the motor efficiency η_{mot}, and the efficiency of the variable-speed drives η_{dr} should all be considered.

The wire-to-water efficiency of a water system η_{ww}, expressed either in dimensionless form or as a percentage, is defined as the ratio of energy output from water to the energy input to the electric wire connected to the motor. It can be calculated as

$$\eta_{ww} = \eta_p \eta_{dr} \eta_{mot} \tag{13-5}$$

The total efficiency of the centrifugal pump η_p can be obtained from the pump manufacturer or calculated from Equation. (13-4). The pump efficiency η_p depends on the type and size of pump as well as the percentage of design volume flow rate during operation. Pump efficiency usually varies from 0. 7 to 0. 85 at the design volume flow rate. Drive efficiency η_{dr} indicates the efficiency of a direct drive, belt drive, and various types of variable-speed drives. For direct drive, $\eta_{dr} = 1$. Among variable-speed drives, an adjustable-frequency alternating-current (ac) drive has the highest drive efficiency. For a 25 • hp (18. 7 • kW) motor, η_{dr} often varies from 0. 96 at design flow to 0. 94 at 30 percent design flow to 0. 80 at 20 percent design flow. Motor efficiency η_{mot} depends on the type and size of motor. It normally varies from 0. 91 for a 10 • hp (7. 5 • kW) high-efficiency motor to 0. 96 for a 250 • hp (187 • kW).

Pump Laws

The performance of geometrically and dynamically similar pump-piping systems 1 and 2 can be expressed as follows:

$$\frac{\dot{V}_2}{\dot{V}_1} = \frac{D_2^3 n_2}{D_1^3 n_1} \tag{13-6a}$$

$$\frac{\Delta H_{t2}}{\Delta H_{t1}} = \frac{n_2^2}{n_1^2} \tag{13-6b}$$

$$\frac{P_2}{P_1}=\frac{n_2^3}{n_1^3} \tag{13-6c}$$

where $\dot{V}$—volume flow rate of pump-piping system, gpm (m^3/s)

ΔH_t—total head lift, ft WC (m WC)

P—pump power input at shaft, hp (kW)

D—outside diameter of pump impeller, ft (m)

n—speed of pump impeller, rpm

Equations(13-6a) through (13-6c) are known as the pump laws. They are similar to the fan laws and are discussed in detail in Unit 12.

13.2 Performance and system Curves

Pump Performance Curves

Pump performance is often illustrated by a head-capacity $H_t-\dot{V}_p$ curve and a power-capacity $P_p-\dot{V}_p$ curve, as shown in Figure 13-3. The head-capacity curve illustrates the performance of a centrifugal pump from maximum volume flow to the shutoff point. If the total head at shutoff point H_{so} is 1.1 to 1.2 times the total head at the point of maximum efficiency H_{ef}, the pump is said to have a flat head-capacity curve. If $H_{so}>1.1\ H_{ef}$, it is a steep-curve pump.

Head pressure is measured in feet of head or psi, and flow is measured in gpm. When a pump is required to pump water higher and higher, the head pressure will increase. The head pressure required to deliver water from the basement to the third floor will be much higher than the head pressure required for delivering water from the basement to the first floor.

As the head pressure increases, the total flow (gpm) through the pump decreases, as shown on the pump performance curve in Figure 13-4. This pump would be able to

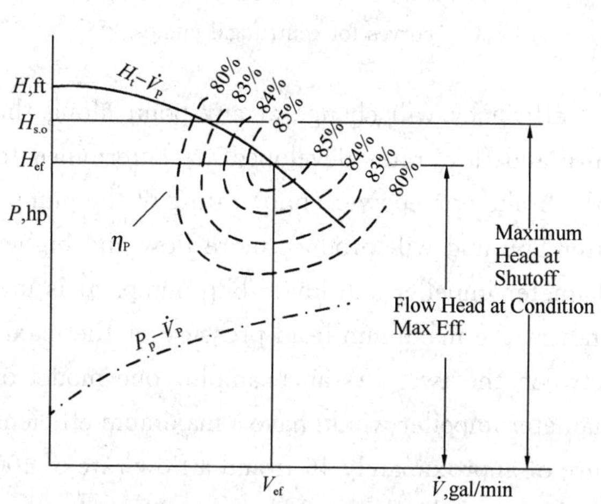

Figure 13-3 Performance curves for centrifugal pumps.

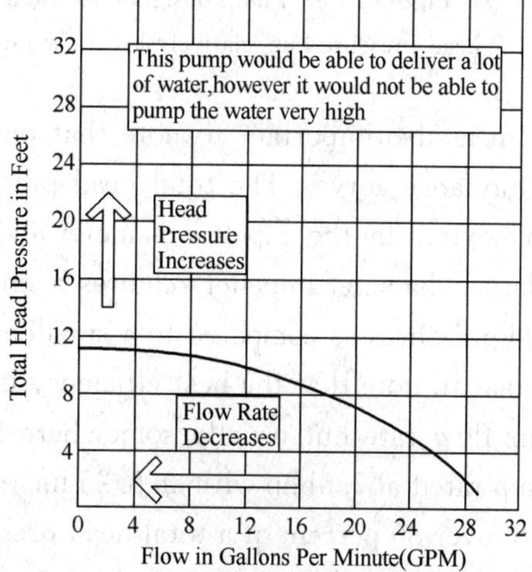

Figure 13-4 As head pressure increases, the pump flow rate decreases.

deliver approximately 18 to 19 gpm with 8 ft of total head. Some pumps such as this are designed for high flow rates and low head pressures. These are ideal for applications where a lot of water is required that does not have to travel too far or too high. Other pumps are designed with high head pressures but lower flow rates, as shown on the pump performance curve in Figure 13-5. These are ideal for applications where less flow is required but the water has to be delivered to the second or third floor of the building.

Pumps are selected to perform a specific purpose, providing the proper flow at a pressure to overcome the resistance of the circuit in which they are placed. Manufacturers provide pump design data, and it is important to select the proper pump depending on the application. An example of a performance chart for an assortment of circulator pumps is shown in Figure 13-6. This allows for matching the most suitable pump to meet the system requirements.

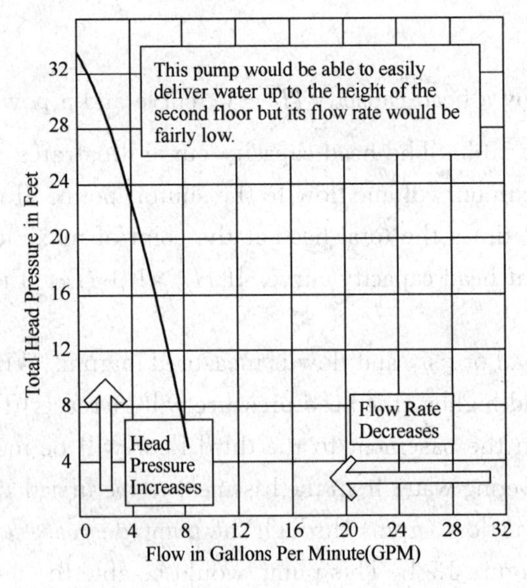

Figure 13-5 Pump designed for high head pressure but relatively low flow rate.

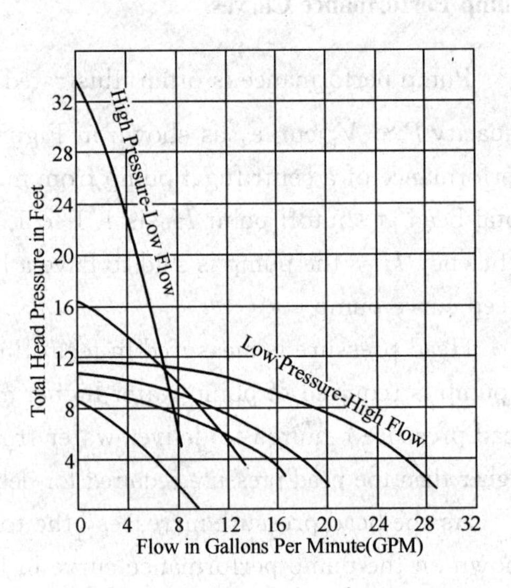

Figure 13-6 Typical performance curves for centrifugal pumps.

It is also important to note that pump efficiency will change at any point along the performance curve. The total head pressure and flow rates developed are determined to some extent by the impeller diameter and the brake horsepower (bhp) rating of the pump. A larger-diameter impeller requires a greater bhp and will produce more flow and higher head pressures as compared to a smaller-diameter impeller and lower bhp pump. It is important to note that the best efficiency is not at the maximum head pressure or the maximum flow rate but usually somewhere between the two. As an example, one model of pump rated at 7. 5bhp with an 8. 25 inch diameter impeller would have a maximum efficiency of over 80 percent at a total head pressure of approximately 46 ft and a flow rate of 400 gpm. If the total head pressure increases to 50 ft and the flow rate drops to 325 gpm, then the efficiency for this pump would drop to less than 77 percent. The same is true if the total head decreases and the flow rate increases.

System Curves

When a pump is connected with a pipe system, it forms a pump-piping system. A water system may consist of one pump-piping system or a combination of several pump-piping systems.

The speed of a variable-speed pump in a variable-flow water system is often controlled by a pressure-differential transmitter installed at the end of the supply main, with a set point normally between 15 and 20 ft WC (4.5 and 6 m WC). This represents the head loss resulting from the control valve, pipe fittings, and pipe friction between the supply and return mains at the farthest branch circuit from the variable-speed pump. Therefore, the head losses of a pump-piping system can be divided into two parts:

- Constant, or fixed, head losso H_{fix}, which remains constant as the water flow varies. Its magnitude is equal to the set point of the pressure-differential transmitter ΔH_{set}, or the head difference between the suction and the discharge levels of the pump in open systems ΔH_{sd}[ft WC (m WC)].
- Variable head loss a H_{var}, which varies as the water flow changes. Its magnitude is the sum of the head losses caused by pipe friction ΔH_{pipe}, pipe fittings ΔH_{fit}, equipment ΔH_{eq}(such as the pressure drop through the evaporator, condenser, and coils), and components ΔH_{cp}, all in ft WC (m WC), that is,

$$\Delta H_{var} = \Delta H_{pipe} + \Delta H_{fit} + \Delta H_{eq} + \Delta H_{cp} \quad (13\text{-}7)$$

Head losses ΔH_{fix} and ΔH_{var} are shown in Figure 13-7. The relationship between the pressure loss p [ft WC (kPa)]; flow head Hvar [ft WC (m WC)]; flow resistance of the water system R_{var}[ft WC/(gpm)^2 (m WC · s^2/m^6)]; and water volume flow rate $\dot{V}_w$ [gpm (m^3/s)], can be expressed as

$$\Delta H_{var} = \Delta p \frac{g_c}{\rho_w g}$$

$$\Delta p = R_{var} \dot{V}_w^2 \quad (13\text{-}8)$$

$$\Delta H_{var} = R'_{var} \dot{V}_w^2$$

where ρ_w——density of water, lb/ft^3 (kg/m^3)

g——gravitational acceleration, ft/s^2 (m/s^2)

g_c——dimensional constant, 32.2 lb_m · ft/lbf · s^2

The curve that indicates the relationship between the flow head, flow resistance, and water volume flow rate is called the *system curve* of a pump-piping system, or a water system.

System Operating Point

The intersection of the pump performance curve and the water system curve is the system operating point of this variable-flow water system, as shown by point P in Figure

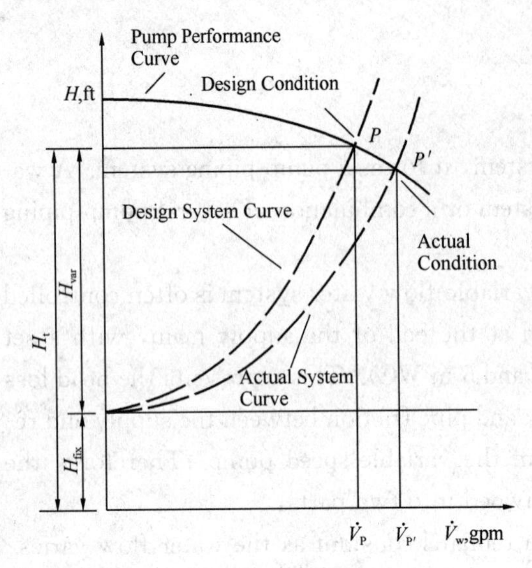

Figure 13-7 Water system curve and system operating point.

13-7. Its volume flow rate is represented by $\dot{V}_P$ [gpm (m³/s)], and its total head is $\Delta H_P = \Delta H_{fix} + xH_{var}$[ft WC (m WC)].

Usually, the calculated system head loss is overestimated, and the selected pump is oversized with a higher pump head, so that the actual system operation point is at point P'. Therefore, for a variable-flow water system installed with a constant-speed pump, the design system operating point is preferably located to the left of the region of pump maximum efficiency, because the system operating point of an oversized pump moves into or nearer to the region of pump maximum efficiency.

Combination of Pump-Piping Systems

When two pump-piping systems 1 and 2 are connected in series as shown in Figure 13-8a, the volume flow rate of the combined pump-piping system, $\dot{V}_{com}$ [gpm (m³/s)] is

$$\dot{V}_{com} = \dot{V}_1 + \dot{V}_2 \tag{13-9}$$

where $\dot{V}_1$ and $\dot{V}_2$ are the volume flow rate of pump-piping systems 1 and 2, gpm (m³/s). The total head lift of the combined system ΔH_{com}[ft WC (m WC)] is

$$\Delta H_{com} = \Delta H_1 + \Delta H_2 \tag{13-10}$$

where ΔH_1 and ΔH_2 are the head of pump-piping systems 1 and 2, ft WC (m WC).

It is simpler to use one system curve to represent the whole system, i. e. , to use a combined system curve. The system operating point of the combined pump-piping system is illustrated by point P with a volume flow of $\dot{V}_P$ and head of ΔH_P. The purpose of connecting pump-piping systems in series is to increase the system head.

When a pump-piping system has parallel-connected water pumps, its volume flow rate [gpm (m³/s)] is the sum of the volume flow rates of the constituent pumps $\dot{V}_1$ $\dot{V}_2$, etc. The head of each constituent pump and the head of the combined pump-piping system are equal. It is more convenient to draw a combined pump curve and one system curve to determine their intersection, the system operating point P, as shown in Figure 13-8*b*. The purpose of equipping a water system with parallel-connected water pumps is to increase its volume flow rate.

Modulation of Pump-Piping Systems

Modulation of the volume flow rate of a pump-piping system can be done by means of the following:

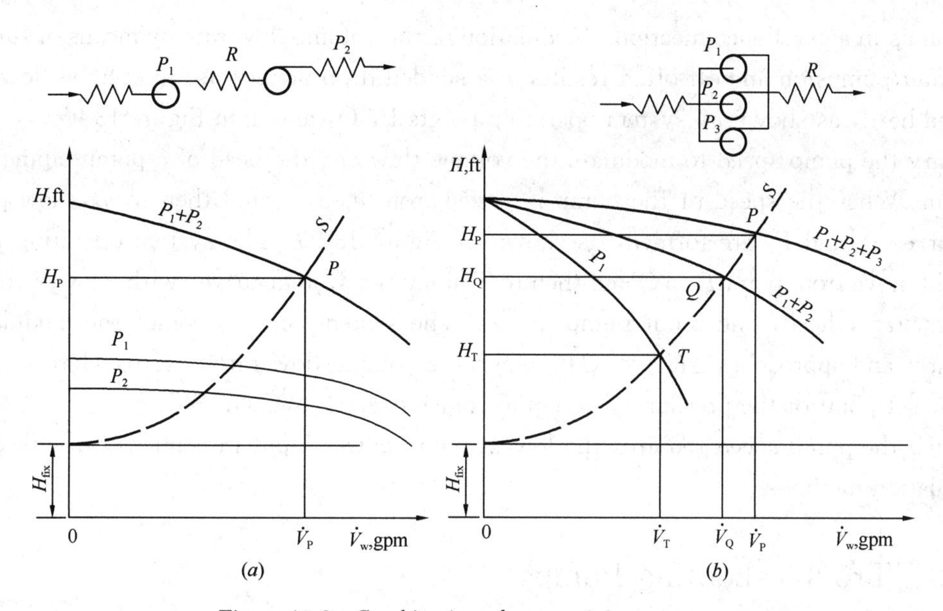

Figure 13-8 Combination of pump-piping systems.

(*a*) Two pump-piping systems connected in series; (*b*) Three parallel-connected pumps

- Throttle the volume flow by using a valve. As the valve closes its opening, the flow resistance of the pump-piping system increases. A new system curve is formed, which results in having a new system operating point that moves along the pump curve to the left-hand side of the original curve, with a lower volume flow rate and higher total head, as shown in Figure 13-9*a*. Such behavior is known as riding on the curve. Using the valve to modulate the volume flow rate of a pump-piping system always wastes energy because of the head loss across the valves ΔH_{val} in Figure 13-9*a*.
- Turn water pumps on or off in sequence for pump-piping systems that have multiple

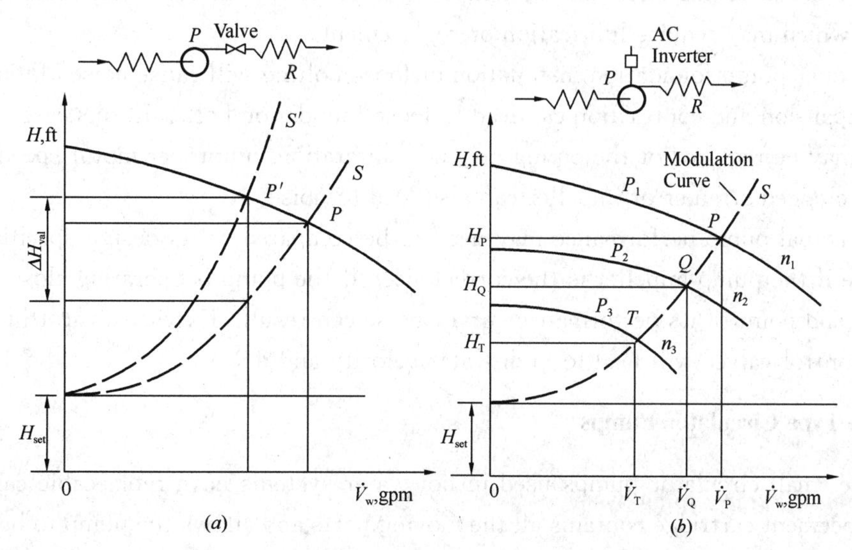

Figure 13-9 Modulation of pump-piping systems.

(*a*) Using a valve; (*b*) Varying the pump speed

pumps in a parallel connection. Modulation of the volume flow rate by means of turning water pumps on and off often results in a sudden drop or increase in volume flow rate and head, as shown by system operating points P, Q, and T in Figure 13-9*b*.

- Vary the pump speed to modulate the volume flow and the head of a pump-piping system. When the speed of the pump is varied from n_1 to n_2 and then to n_3, new pump curves P_2 and P_3 are formed, as shown in Figure 13-9*b*. The system operating point will move from point P to Q and then to T along the system curve, with a lower volume flow rate, head, and input pump power. The system curve becomes the modulating curve and approaches $\Delta H_{fix} = xH_{set}$ when the volume flow rate is zero. Here ΔH_{set} is the set point of the pressure-differential transmitter, ft WC (m WC).

Varying the pump speed requires the lowest pump power input in comparison with other modulation methods.

13. 3 Troubleshooting Pumps

A common troubleshooting problem is pump noise and vibration. This is a nuisance and the vibration can cause the piping to fatigue and break over time. If the pump loses its prime, its impeller can overheat and become so damaged that the pump will not work. Common causes of pump noise and vibration are misalignment, air entrainment, and clogged strainers.

Pump or System Noise

Noise could be the result of a shaft misalignment which would need to be checked and then realigned if necessary. A worn coupling can lead to noise and if this is the cause it will need to be replaced and the shaft realigned. The same is true for worn pump or motor bearings which may require lubrication or replacement.

Improper pump foundation installation or loose bolting will cause noise. Piping strain due to expansion and contraction can lead to loose foundation bolts. In this case additional hangers may be required or the piping may need alteration. Improper motor speed or rotation or a clogged strainer or impeller can also lead to noise.

The actual pump performance may need to be compared to its design specification to determine if the pump impeller is the correct size. If the pump is operating close to or beyond the end point of its performance curve, noise can result. Excessive throttling of balance or control valves will lead to high water velocity and noise.

Cartridge-Type Circulator Pumps

Some small circulator pumps used in hot-water systems have replaceable cartridges. The replacement cartridge contains all the moving parts and allows the pump to be serviced instead of replacing the entire unit. It is self-lubricating and contains no mechanical seal. When troubleshooting this type of pump, the motor windings can be easily inspected. Notice the charred burnt windings of the motor and the discoloration of the capacitor.

Inadequate or No Circulation

This is often the result of an improperly filled system that is air bound. Typically the vent piping, make up feed regulator, or expansion tank may be at fault. It is important to get all of the air out of the piping system before leaving the job. Air trapped in the system can cause noise and may result in the water pump losing its prime. Radiant piping systems will often have automatic venting valves located at the high points of the circulating water loop. If it is determined that air is not the problem, then another possible cause is a clogged suction strainer.

Circulating pumps on closed-loop systems only work against the vertical lift head when the system is initially being filled. Once the system is completely filled with water, the pump effective head (discharge head - suction head) is only the flow resistance of the piping system itself. The suction head developed from a height of 10 ft of water on the return side of the pump will balance the discharge head developed from a height of 10 ft of water on the outlet side of the pump.

Unit 14 Refrigerant

A refrigerant is the primary working fluid used for absorbing and transmitting heat in a refrigeration system. Refrigerants absorb heat at a low temperature and low pressure and release heat at a higher temperature and pressure. Most refrigerants undergo phase changes during heat absorption—evaporation and heat releasing—condensation. A much larger amount of heat can be absorbed and transferred when a liquid is vaporized to a gas. This method takes advantage of the latent heat characteristics of the refrigerant and results in much higher operating efficiency due to the reduced quantity of refrigerant that must be circulated. Refrigerants in mechanical refrigeration systems do this by changing their physical state. They absorb heat by evaporating from a liquid to a vapor; they release heat by condensing from a vapor back to a liquid. The refrigerant's temperature is controlled by its pressure. Lower pressures yield lower temperatures, while higher pressures yield higher temperatures.

Any study of the refrigeration cycle would be incomplete without a look at the refrigerant in the system. The refrigerant is the first component of a refrigeration system that must be determined, since its chemical and physical properties will play a large role in selecting all other components in a refrigeration system. A number of materials have physical and thermodynamic properties that make them suitable for use as refrigerants. In fact, anything that can boil at a low temperature and condense at a higher temperature can be used as a refrigerant. Water can even be used as a refrigerant under the right conditions. Although water boils at 212℉ at standard atmospheric pressure, it boils at 40℉ in a vacuum of 29.75 in Hg.

14.1 Properties of Refrigerants

Refrigerants have many physical and chemical characteristics that determine their suitability for any particular system. Today, the preservation of the ozone layer is the first priority of refrigerant selection. In addition, the global warming effect and the following factors should be considered.

Safety Requirements

Refrigerant may leak from pipe joints, seals, or component parts during installation, operation, or accident. Therefore, refrigerants must be acceptably safe for humans and manufacturing processes, with little or no toxicity or flammability.

In ANSI/ASHRAE Standard 34-1997, the toxicity of refrigerants is classified as class

A or B. Class A refrigerants are of lower toxicity. A class A refrigerant is one whose toxicity has not been identified when its concentration is less than or equal to 400ppm, based on threshold limit value-time-weighted average (TLV-TWA) or equivalent indices. The TLV-TWA concentration is a concentration to which workers can be exposed over an 8-h workday and a 40-h workweek without suffering adverse effect. Concentration ppm means parts per million by mass.

Class B refrigerants are of high toxicity. A class B refrigerant produces evidence of toxicity when workers are exposed to a concentration below 400ppm based on a TLV-TWA concentration. Flammable refrigerants explode when ignited. If a flammable refrigerant is leaked in the area of a fire, the result is an immediate explosion. Soldering and welding for installation or repair cannot be performed near such gases.

ANSI/ASHRAE Standard 34-1997 classifies the flammability of refrigerants into classes 1, 2, and 3. Class 1 refrigerants show no flame propagation when tested in air at a pressure of 14.7psia (101kPa) at 65°F (18.3℃). Class 2 refrigerants have a lower flammability limit (LFL) of more than 0.00625 lb/ft^3 (0.1kg/m^3) at 70°F (21.1℃) and 14.7 psia (101kPa abs.), and a heat of combustion less than 8174 Btu/lb (19,000kJ/kg). Class 3 refrigerants are highly flammable, with an LFL less than or equal to 0.00625 lb/ft^3 (0.1kg/m^3) at 70°F (21.1℃) and 14.7 psia (101kPa abs.) or a heat of combustion greater than or equal to 8174 Btu /lb (19000kJ/kg).

A refrigerant's safety classification is its combination of toxicity and flammability. According to

ANSI/ASHRAE Standard 34-1997, safety groups are classified as follows:

- A1 lower toxicity and no flame propagation
- A2 lower toxicity and lower flammability
- A3 lower toxicity and higher flammability
- B1 higher toxicity and no flame propagation
- B2 higher toxicity and lower flammability
- B3 higher toxicity and higher flammability

Forzeotropic blends whose flammability and toxicity may change as their composition changes, a dual safety classification should be determined. The first classification denotes the classification of the formulated composition of the blend. The second classification lists the classification of the

blend composition at the worst case of fractionation.

Effectiveness of Refrigeration Cycle

The effectiveness of refrigeration cycles, or coefficient of performance (**COP**), is one parameter that affects the efficiency and energy consumption of the refrigeration system. It will be clearly defined in a later section. The COP of a refrigeration cycle using a specific refrigerant depends mainly upon the isentropic work input to the compressor at a given condensing and evaporating pressure differential, as well as the refrigeration effect produced.

Evaporating and Condensing Pressures

It is best to use a refrigerant whose evaporating pressure is higher than that of the atmosphere so that air and other noncondensable gases will not leak into the system and increase the condensing pressure. The condensing pressure should be low because high condensing pressure necessitates heavier construction of the compressor, piping, condenser, and other components. In addition, a high-speed centrifugal compressor may be required to produce a high condensing pressure.

Oil Miscibility

When a small amount of oil is mixed with refrigerant, the mixture helps to lubricate the moving parts of a compressor. Oil should be returned to the compressor from the condenser, evaporator, accessories, and piping, in order to provide continuous lubrication. On the other hand, refrigerant can dilute oil, weakening its lubricating effect; and when the oil adheres to the tubes in the evaporator or condenser, it forms film that reduces the rate of heat transfer.

Inertness

An inert refrigerant does not react chemically with other materials, thus avoiding corrosion, erosion, or damage to the components in the refrigerant circuit.

Thermal Conductivity

The thermal conductivity of a refrigerant is closely related to the efficiency of heat transfer in the evaporator and condenser of a refrigeration system. Refrigerant always has a lower thermal conductivity in its vapor state than in its liquid state. High thermal conductivity results in higher heat transfer in heat exchangers.

Refrigeration Capacity

The cubic feet per minute (cfm) suction vapor of refrigerant required to produce 1 ton of refrigeration (liters per second to produce 1kW of refrigeration) depends mainly on the latent heat of vaporization of the refrigerant and the specific volume at the suction pressure. It directly affects the size and compactness of the compressor and is one of the criteria for refrigerant selection.

Physical Properties

Discharge Temperature. A discharge temperature lower than 212℉(100℃) is preferable because temperatures higher than 300℉(150℃) may carbonize lubricating oil or damage some of the components.

Dielectric Properties. Dielectric properties are important for those refrigerants that will be in direct contact with the windings of the motor (such as refrigerants used to cool the motor windings in a hermetically sealed compressor and motor assembly).

Operating Characteristics

Leakage Detection. Refrigerant leakage should be easily detected. If it is not, gradual capacity reduction and eventual failure to provide the required cooling will result. Most of the currently used refrigerants are colorless and odorless. Leakage of refrigerant from the refrigeration system is often detected by the following methods:

Halide torch. This method is simple and fast. When air flows over a copper element heated by a methyl alcohol flame, the vapor of halogenated refrigerant decomposes and changes the color of the flame (green for a small leak, bluish with a reddish top for a large leak).

- Electronic detector. This type of detector reveals a variation of electric current due to ionization of decomposed refrigerant between two oppositely charged electrodes. It is sensitive, but cannot be used where the ambient air contains explosive or flammable vapors.
- Bubble method. A solution of soup or detergent is brushed over the seals and joints where leakage is suspected, producing bubbles that can be easily detected.

14.2 Classification of Refrigerants

Refrigerants can be divided into five major chemical categories based on their chemical composition:

- Hydrocarbons
- Chlorofluorocarbons
- Hydrofluorocarbons
- Hydrochlorofluorocarbons
- Natural refrigerants

Refrigerants are now frequently referred to by their chemical family using the first letters of each chemical component in the refrigerant to produce abbreviations (HC, CFC, etc.). Common refrigerant chemical families are shown inTable 14-1.

Refrigerant chemical families. **Table 14-1**

Symbol	Refrigerant Family	Chemical Components
HCs	Hydrocarbons	Hydrogen and carbon
CFCs	Chlorofluorocarbons	Chlorine, fluorine, carbon
HFCs	Hydrofluorocarbons	Hydrogen, fluorine, carbon
HCFCs	Hydrochlorofluorocarbons	Hydrogen, chlorine, fluorine, carbon
CO_2, NH_3	Natural refrigerants	Carbon, oxygen, nitrogen, hydrogen

Until recently, most refrigerants were compounds. Water is a good example of a compound. Take a highly flammable gas, hydrogen, combine it chemically with oxygen, a gas that is necessary for combustion, and the result is water, a chemical that does not burn and is widely used to extinguish flames. Two chemicals have been combined to make a new chemical with distinct chemical properties that are completely separate from the properties of the original chemicals. Further, these chemicals cannot be separated by physical means. The hydrogen cannot be distilled from the oxygen by boiling the water. Any pure chemical or compound has a specific saturation temperature for any given pressure. The boiling temperature can be precisely determined if the pressure on the liquid is known. Examples of refrigerants that are compounds include some of the old standards: R-11, R-12, and R-22. These are all based on the same molecule, methane.

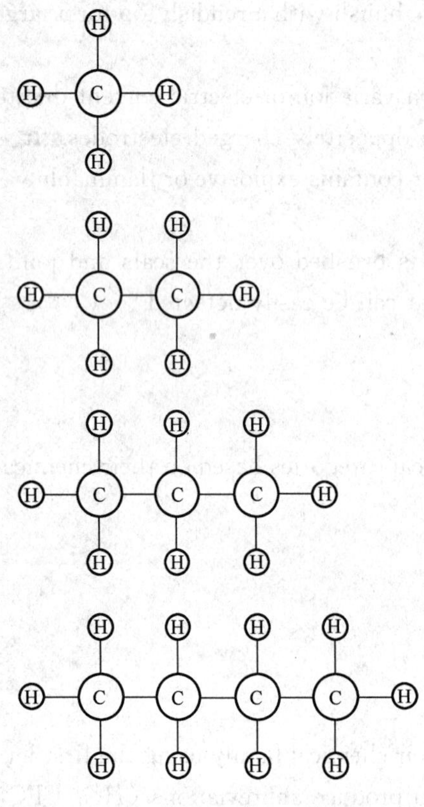

Figure 14-1 Hydrocarbon refrigerants.

HC Refrigerants

Hydrocarbons are compounds consisting of carbon atoms tied to each other in chains surrounded by hydrogen atoms. Figure 14-1 shows the arrangement of hydrogen and carbon atoms in several common hydrocarbons. Normally, hydrocarbons are thought of as fuels; they are burned for energy. Hydrocarbons make excellent refrigerants. In addition to being excellent refrigerants, they are environmentally friendly. Hydrocarbons have zero ozone depletion because they contain no chlorine or bromine, and their global-warming potential is very low compared to halogenated refrigerants. Unfortunately, hydrocarbons are not just flammable but explosive. Hydrocarbons are only approved for limited applications in the United States. They are more widely used in other places, including Australia, Canada, China, and Europe. Hydrocarbons are sometimes used in small quantities in zeotropic blends to help with oil return. The amount of hydrocarbons used in these blends is so low that the mixture retains a nonflammable safety rating. Table 14-2 shows a list of common hydrocarbon compounds and their refrigerant number.

Refrigerant chemical families. **Table 14-2**

Refrigerant Number	Refrigerant Name	No. Hydrogen Atoms	No. Carbon Atoms
R-50	Methane	4	1
R-170	Ethane	6	2
R-290	Propane	8	3
R-600	Butane	10	4

Halogenated Refrigerants

All halogenated refrigerants are based on hydrocarbons, like methane. Halogenated refrigerants are formed by combining a hydrocarbon with halogens. Halogens are a group of five highly reactive chemicals:

- Chlorine
- Fluorine
- Bromine
- Iodine
- Astatine

Chlorine and fluorine are the primary halogens used to make halogenated refrigerants. The halogens replace hydrogen atoms in the hydrocarbon molecules. If only some of the hydrogen atoms are replaced, the refrigerant is partially halogenated. If all of the hydrogen atoms are replaced, the refrigerant is fully halogenated. Fully halogenated CFC refrigerants are more chemically stable than partially halogenated HCFC and HFC refrigerants. Partially halogenated HCFC and HFC refrigerants can be more easily broken down because of the presence of hydrogen. Some advantages of halogenated refrigerants are that they are generally nonflammable, low in toxicity, and do not chemically react with many materials. This is primarily because they are very chemically stable molecules. The chlorine in halogenated refrigerants and the bromine in halon fire suppressant are the halogens responsible for ozone depletion. Chemicals that contain either chlorine or bromine contribute to ozone depletion.

CFC Refrigerants

CFC is short for chlorofluorocarbon. CFC refrigerants contain chlorine, fluorine, and carbon. They are fully halogenated refrigerants with chlorine and fluorine replacing all the hydrogen atoms of a hydrocarbon compound. Most CFC refrigerants are built around the methane or ethane molecules. Figure 14-2 shows the chemical structure of R-12, which is built on a methane molecule.

CFC refrigerants were once the most common type of refrigerants in use. However, CFC refrigerants are the worst offenders in terms of ozone depletion; they have the highest ozone-depletion potentials of any group of refrigerants. CFCs also have a relatively high global-warming index and a long atmospheric lifetime. CFCs make good refrigerants because they are very chemically stable. This stability is also what makes them an environmental liability. CFCs do not break down in lower earth atmosphere. They find their way up to the stratosphere, where the high-intensity ultraviolet light breaks the CFCs apart, releasing the chlorine in the stratosphere.

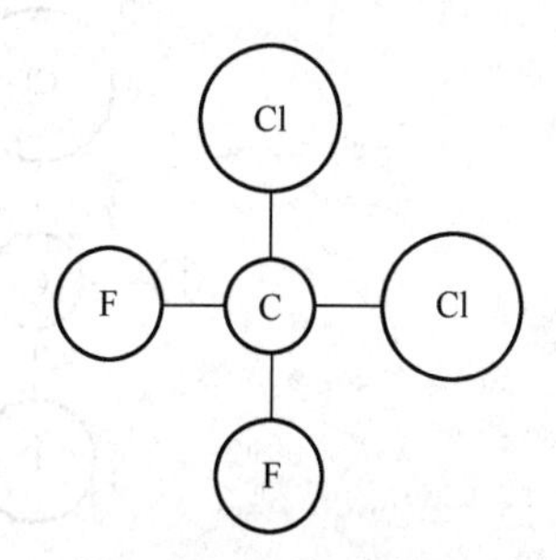

Figure 14-2 CFC-12 is based on the methane molecule.

It has been illegal to intentionally vent CFC refrigerants into the atmosphere since July, 1, 1992, and 1995 was the last year that CFCs were allowed to be manufactured or imported into the United States, so the supply of CFCs is very low. Some CFC refrigerant is still available at very high prices, but there is now very little demand. Table 14-3 lists some of the most common CFC refrigerants.

CFC refrigerants. Table 14-3

Refrigerant Number	Refrigerant Name	No. Hydrogen Atoms	No. Chlorine Atoms	No. Fluorine Atoms
R-11	Trichlorofluoromethane	1	3	1
R-12	Dichlorodifluoromethane	1	2	2
R-13	Chlorotrifluoromethane	1	1	3

HCFC Refrigerants

HCFC is short for hydrochlorfluorocarbon. HCFCs contain hydrogen, chlorine, fluorine, and carbon. Most HCFC refrigerants are built around the methane or ethane molecules. Figure 14-3 shows the chemical structure of R-22, which is built on a methane molecule. HCFCs are only partially halogenated because they still have some hydrogen atoms. This makes them less stable than CFC refrigerants and more environmentally friendly. HCFCs still have an ozone-depletion potential because they still have chlorine. Their ozone-depletion potential is far less than the ozone-depletion potential of CFC refrigerants because HCFCs are more likely to break down in the lower atmosphere. Once the chlorine is released in the lower atmosphere it finds plenty of chemicals to attack, so it will not reach the stratosphere. The most common HCFC refrigerant is R-22. Since January 1, 2010, new equipment charged with R-22 cannot be manufactured or imported, and R-22 is scheduled for a total phaseout on January 1, 2020. All HCFC refrigerants are scheduled to be phased out on January 1, 2030. R-22 has been the least expensive halogenated refrigerant available, but the price of R-22 is rising quickly because its production has been cut back. Table 14-4 lists some of the most common HCFC refrigerants.

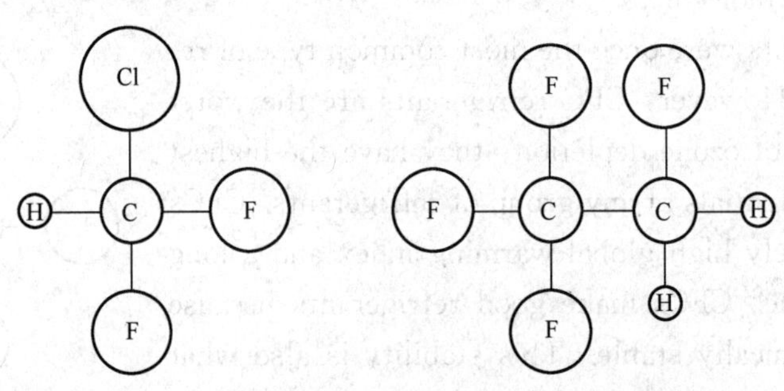

Figure 14-3 HCFC-22 is based on the methane molecule.

Figure 14-4 HFC 134a is based on the ethane molecule.

HCFC refrigerants. **Table 14-4**

Refrigerant Number	Refrigerant Name	No. Carbon Atoms	No. Hydrogen Atoms	No. Chlorine Atoms	No. Fluorine Atoms
R-22	Monochlorodifluoromethane	1	1	1	2
R-123	Dichlorotrifluoroethan	1	1	2	3
R-124	Chlorotetrafluoroethan	2	1	1	4
R-142b	Chlorodifluoroethan	2	3	1	2

HFC Refrigerants

HFC is short for hydrofl uorocarbon. HFCs contain hydrogen, fluorine, and carbon. Most HFC refrigerants are built around the ethane molecule. Figure 14-4 shows the chemical structure of R-134a, which is built on an ethane molecule. HFC refrigerants are only partially halogenated because they still have some hydrogen atoms. This makes them less stable than CFC refrigerants. They are more environmentally friendly than either CFC or HCFC refrigerants because they contain no chlorine and therefore have no ozonedepletion potential. However, HFC refrigerants do still have a global-warming potential. It has been illegal to intentionally vent HFC refrigerants into the atmosphere since November 15, 1995. Table 14-5 lists some of the most common HFC refrigerants.

HFC refrigerants. **Table 14-5**

Refrigerant Number	Refrigerant Name	No. Carbon Atoms	No. Hydrogen Atoms	No. Fluorine Atoms
R-125	Pentafluoroethane	2	1	5
R-134a	Tetrafluoroethane	2	2	4
R-23	Trifluoromethane	1	1	3
R-32	Difluoromethane	1	1	2
R-143a	Trifluoroethane	2	3	3
R-152a	Difluoroethane	2	4	2

CO_2 Refrigerants

Interest in using CO_2 as a refrigerant has been increasing because it is nontoxic and nonflammable, does not deplete the ozone layer, has a very low global-warming potential, and is inexpensive. The biggest challenge is the extremely high pressures necessary for CO_2 systems to operate. A pressure of over 1,400 psig is necessary for an air-cooled system similar to a common refrigeration system. CO_2 will not condense at temperatures above 88°F because the temperature and pressure in the high side exceed the critical point. Heat is rejected by sensible cooling of the gas. These CO_2 systems are called transcritical systems because they operate both above and below the critical point. The high side of a transcritical system operates above the critical point, and the low side operates below the critical point. CO_2 can also be applied as the low-temperature portion of a cascade system, where the CO_2 condenser is cooled by the evaporator of another refrigeration system that uses a more traditional refrigerant. By keeping the condenser at a low temperature, the CO_2 system operates like a "normal" refrigeration system, below the critical point.

All refrigerants are combinations of two or more chemicals, but there are differences in how these chemicals are combined. The chemicals used to form a refrigerant can be combined in one of three ways:

- Compounds
- Zeotropes
- Azeotropes

The terms *zeotropic* and *azeotropic* derive from Greek. Zeo means "to boil", and trope means "to turn". Thus, zeotropic refrigerants turn, or change, as they boil. The mixture changes (trope) as it boils (zeo). The prefix A - means "not". Placing the prefix A in front of a word effectively turns the word upside down. For example, an amoral person is one who does not have morals. An azeotropic refrigerant does not change as it boils.

14.3 Refrigerant Replacement

The use of CFCs and HCFCs is a global concern. Approximately two-thirds of all fully halogenated CFCs were used outside the United States in the mid-1980s. In 1985, the total use of halocarbons in the United States was 611 million lb (0.28 million ton). These halocarbons were used in foam insulation, automotive air conditioners, new systems of Air Conditioning and Refrigeration Institute (ARI) members, and other products. Foam insulation blown by CFCs was the largest user. Automotive air conditioners made up 19 percent of the total and CFCs purchased by ARI members for new systems made up 5 percent of the total use. Of the CFCs and HCFCs purchased by ARI members, HCFC-22 made up 77 percent, while CFC-11 and CFC-12 each made up about 10 percent.

Ozone Depletion and Global Warming Potentials

To compare the relative ozone depletion caused by various refrigerants, an index called the *ozone depletion potential* (**ODP**) has been proposed. ODP is the ratio of the rate of ozone depletion of 1 lb of any halocarbon to that of 1 pound of CFC-11. The ODP of CFC-11 is assigned a value of 1. The ODP of most refrigerants is less than 1, but there are other substances, such as halons, with ODPs much greater than 1. Note that refrigerants that do not contribute to ozone depletion are not classified in this system. The EPA divides ozone-depleting refrigerants into two classes based on their ozone-depletion potential.

- Class I refrigerants have an ODP greater than 0.2.
- Class II refrigerants have an ODP less than 0.2.

This classification is becoming a historical artifact because nearly all Class I refrigerants are CFCs, and CFC production and importation have been banned since the end of 1995. It is still possible to purchase CFC refrigerants from a supplier who still has refrigerant left over from 1995, or from a refrigerantreclaimer. Class II refrigerants will be around for some time to come, but their time is also limited. HCFC refrigerants are scheduled to be phased out completely by 2030. Table 14-6 shows refrigerant classes according to their ozone-depletion potential.

Classes of ozone-depleting substances. **Table 14-6**

Refrigerant Class	Ozone Depletion Potential	Refrigerants
Class I	ODP> 0.2	CFC-11, CFC-12, R-500, R-502
Class II	ODP< 0.2	HCFC-22, HCFC-123

Similar to the ODP, the *halocarbon global warming potential* (**HGWP**) is the ratio of calculated warming for each unit mass of gas emitted to the calculated warming for a unit mass of reference gas CFC-11. The HGWPs of various refrigerants are given in relative handbooks. In addition to the HGWP, another global warming index uses CO_2 as a reference gas. For example, 1 lb of HCFC-22 has the same effect on global warming as 4100lb (1860kg) of CO_2 in the first 20 years after it is released into the atmosphere. Its impact drops to 1500lb (680kg) at 100 years.

Phaseout of CFCs, Halons, and HCFCs

The theory of depletion of the ozone layer was proposed in 1974 by F. S. Rowland and M. J. Molina. (The 1995 Nobel Prize was awarded to F. Sherwood Rowland, Mario Molina, and Paul Crutzen for their work in atmospheric chemistry and theory of ozone depletion.) Network station in Halley Bay, Antarctica, established a baseline trend of ozone levels that helped scientists to discover the ozone hole in 1985. National Aeronautics and Space Administration (NASA) flights into the stratosphere over the arctic and antarctic circles found CFC residue where the ozone layer was damaged. Approximately the same ozone depletion over the antarctic circle was found in 1987, 1989, 1990, and 1991. By 1988, antarctic ozone levels were 30 percent below those of the mid-1970s. The most severe ozone loss over the antarctic was observed in 1992. Ground monitoring at various locations worldwide in the 1980s has showed a 5 to 10 percent increase in ultraviolet radiation. The loss of ozone over Antarctica is referred to as the"ozone hole. "Many people once believed that ozone depletion could only occur at the poles, but reductions in ozone have also been observed in northern middle latitudes. Although there is controversy about the theory of ozone layer depletion among scientists, as discussed in Rowland (1992), action must be taken immediately before it is too late.

Montreal Protocol and Clean Air Act

In 1978, the Environmental Protection Agency (EPA) and the Food and Drug Administration (FDA) of the United States issued regulations to phase out the use of fully halogenated CFCs in nonessential aerosol propellants, one of the major uses at that time. On

September 16, 1987, the European Economic Community and 24 nations, including the United States, signed the Montreal Protocol. This document is an agreement to phase out the production of CFCs and halons by the year 2000. The Montreal Protocol had been ratified by 157 parties.

The Clean Air Act Amendments, signed into law in the United States on November 15, 1990, governed two important issues: thephaseout of CFCs and a ban (effective July 1, 1992) on the deliberate venting of CFCs and HCFCs. Deliberate venting of CFCs and HCFCs must follow the regulations and guidelines of the EPA. In February 1992, then-President Bush called for an accelerated phaseout of the CFCs in the United States. Production of CFCs must cease from January 1, 1996, with limited exceptions for service to certain existing equipment.

In late November 1992, representatives of 93 nations meeting in Copenhagen also agreed to the complete cessation of CFC production beginning January 1, 1996, and of halons by January 1, 1994, except continued use from existing (reclaimed or recycled) stock in developed nations. In addition, the 1992 Copenhagen amendments and later a 1995 Vienna meeting revision agreed to restrict the production of HCFCs relative to a 1989 level beginning from 2004 in developed nations according to the schedule listed in Table 14-7.

Production of HCFCs. **Table 14-7**

Date	Production limit
	100 percent cap
January 1, 1996	Cap = 2.8 percent of ODP of 1989 CFC consumption plus total ODP of 1989 HCFC consumption
January 1, 2004	65 percent cap
January 1, 2010	35 percent cap
January 1, 2015	10 percent cap
January 1, 2020	0.5 percent cap
January 1, 2030	Complete cessation of production

Consumption indicates the production plus imports minus exports and feedstocks. The value of 2.8 percent cap is the revised value of the Vienna meeting in 1995 to replace the original value of 3.1 percent in the Copenhagen amendments. The Copenhagen amendments had been ratified by 58 parties.

Action and Measures

The impact of CFCs on the ozone layer poses a serious threat to human survival. The following measures are essential:

Conversions and Replacements. Use alternative refrigerants (substitutes) to replace the CFCs in existing chillers and direct-expansion (DX) systems. During the conversion of the CFC to non-ozone depletion alternative refrigerants, careful analysis should be conducted of capacity, efficiency, oil miscibility, and compatibility with existing materials after conversion. For many refrigeration systems that already have a service life of more than 15 years, it may be cost-effective to buy a new one using non-CFC refrigerant to replace the existing refrigeration package.

- HFC-134a and HCFC-22 are alternative refrigerants to replace CFC-12.
- HCFC-123, and HFC-245ca are alternative refrigerants to replace CFC-11 in large chillers. It is important to realize that HCFC-123 and HCFC-22 themselves are interim refrigerants and will be restricted in consumption beginning in 2004. HCFC-123 has a very low global warming potential and is widely used in centrifugal chillers. HCFC-22 is widely used in small and medium-size DX systems.
- HFC-134a, HFC-407C, and HFC-410A are alternative refrigerants to replace HCFC-22. HFC-407C is a near-azeotropic refrigerant of HFC-32/HFC-125/HFC-134a (23/25/52) [means (23%/25%/525)], and HFC-410A also a near-azeotropic refrigerant of HFC-32/HFC-125 (50/50).
- HFC-245ca or another new HFC possibly developed before 2004 will be the hopeful alternative to replace HCFC-123.

  In supermarkets, CFC-502 is a blend of HCFC-22/CFC-115 (48.8/51.2).
- HFC-404A, HFC-507, and HFC-410A are alternative refrigerants to replace CFC-502. HFC- 404A s a near-azeotropic refrigerant of HFC-125/HFC-143a/HFC-134a(44/52/4); and HFC-507 is an azeotropic refrigerant of HFC-125/HFC-143a (45/55).

Reducing Leakage and Preventing Deliberate Venting. To reduce the leakage of refrigerant from joints and rupture of the refrigeration system, one must detect the possible leakage, tighten the chillers, improve the quality of sealing material, and implement preventive maintenance.

Prevent the deliberate venting of CFCs and HCFCs and other refrigerants during manufacturing, installation, operation, service, and disposal of the products using refrigerants. Avoid CFC and HCFC emissions through recovery, recycle, and reclaiming. According to ASHRAE Guideline 3-1990, recovery is the removal of refrigerant from a system and storage in an external container. Recycle involves cleaning the refrigerant for reuse by means of an oil separator and filter dryer. In reclamation, refrigerant is reprocessed for new product specifications.

To avoid the venting of CFCs and HCFCs and other refrigerants, an important step is to use an ARI-certified, portable refrigerant recovering/recycling unit to recover all the liquid and remaining vapor from a chiller or other refrigeration system. An outside recovery/reclaiming service firm may also be employed.

In addition to the recovery of refrigerants from the chiller or other refrigeration system, refrigerant vapor detectors should be installed at locations where refrigerant from a leak is likely to concentrate. These detectors can set off an alarm to notify the operator to seal the leak.

All replacement refrigerants must be evaluated and approved by the EPA. The program that evaluates new refrigerants is called the Significant New Alternatives Policy, or

SNAP. Substitutes are reviewed on the basis of ozonedepletion potential, global-warming potential, toxicity, flammability, and exposure potential. Lists of acceptable and unacceptable substitutes are updated several times each year.

A "drop in" refrigerant can replace the original system refrigerant without any changes to the system, such as compressor oil changes, metering device adjustments, or seal changes. Although there are many advertised "drop in" replacement refrigerants, the EPA does not recognize any refrigerants as being true drop-in refrigerants requiring no system modifications or servicing.

The EPA lists alternative refrigerants as suitable for retrofit, new application, or unacceptable. Alternatives listed as retrofit refrigerants can be used to replace the refrigerant in an existing system. This does not mean that they will work with no system alterations, but they can be made to work in an existing system. Refrigerants listed as new alternatives will only work in new applications that accomplish the same thing. Sometimes these new alternatives are completely different approaches to cooling. One example is evaporative cooling, which is listed as an alternative for many types of systems. Evaporative cooling is a cooling system that reduces air temperature by evaporating water. Obviously, an evaporative system is a completely different type of system from a traditional mechanical refrigeration system. It is only a replacement in the sense that it accomplishes roughly the same thing as a mechanical refrigeration system if humidity control is not an issue.

There are some refrigerants that are specifically designed to replace CFCs or HCFCs in new equipment. HFC- 134a replaces CFC-12 in new equipment but is not generally considered a retrofit refrigerant. HFC-407C is a very close replacement for HCFC-22 in retrofit situations. HFC-410A is the replacement for HCFC-22 in new equipment. It has pressures 40-60 percent higher than R-22, so it cannot be used as a retrofit. It is a replacement only in the sense that it is used in the same general types of cooling applications where HCFC-22 is used.

There are many refrigerants available for retrofitting HCFC 22 systems, including R-407A, R-407C, R-417A, R-421A, R-422B, R-422C, R-424A, R-427A, and R-437A. However, using a new chemical in a refrigeration system will generally void the equipment's third-party certification, such as UL or CSA, because the system was not tested by the testing organization with the new replacement refrigerant.

Unit 15　System Controls

The purpose of HVACR system is to start, stop or regulate the flow of air, water or steam and to provide stable operation of the system by maintaining the desired temperature, humidity, and pressure. The automatic control system is a group of components, each with a definite function designed to interact with the other components so that the system is self-regulating. HVACR control systems are classified according to the source of power used for the operation of the various components. The classifications and power sources are:

Direct Digital Control (DDC) electronic systems

Low amperage, 4 to 20 milliamps dc (direct current)

Low voltage, 0 to 15 volts dc (vdc)

Electrical Controls

Low voltage, 24 volts ac (alternating current)

Line voltage, 110 to 220 volts ac (vac)

Lighting, 120, 177vac

Motor, 208, 220-240, 440-480vac

Pneumatic Controls using compressed air

Interface Controls using electronic, electrical and pneumatic components

Sensors, Controllers, Relays and Switches, Actuators or Motors, and Controlled Devices

15.1　Direct Digital Control Systems

Direct Digital Control (DDC) is the automated control of a condition or process by a digital computer. A comparison can be made between conventional pneumatic controls and DDC. A typical HVACR pneumatic control system may consist of a pneumatic temperature sensor, pneumatic controller, and a heating or cooling valve. In the pneumatic system the sensor provides a signal to the controller and the controller provides an output to the valve to position it to provide the correct temperature of supply air. A DDC system replaces this local control loop with an electronic temperature sensor and a microprocessor to replace the controller. The output from the microprocessor is converted to a pressure signal to position the same pneumatic heating and cooling valve as in the pneumatic control system. However, the DDC system is not limited to utilizing pneumatic control devices but may also interface with electric or electronic actuators.

Electric or pneumatic devices can be used to provide the control power to the final control elements (the controlled device), but the DDC system provides the signal to that device. In a true DDC system there is no conventional controller. The controller has been replaced by the microprocessor. A common application of DDC includes the control of the heating valve, cooling valve, mixed air damper, outside air damper, return air damper, and economizer cycles to maintain the desired supply air temperature. Other systems commonly controlled by DDC include: chilled water temperature, hot water temperature, and variable air volume and variable water volume capacity.

The DDC system uses a combination of software algorithms (mathematical equations) and hardware components to maintain the controlled variable according to the desires of the system operator. The controlled variables may be temperature, pressure, relative humidity, etc. In the past, the maintenance and operations personnel had to calibrate the local loop controller at the controller's location. Now, with a DDC system, the system's operator may tune the control loop by changing the software variables in the computer using the operator's keyboard. So, instead of calibrating the hardware controller the control sequence and setpoint are input to the computer by a software program and modified by a proper password and the appropriate command keyboard entry.

The DDC system monitors the controller variable and compares its value to the desired value stored in the computer. If the measured value is less than or greater than the desired value, the system output is modified to provide the correct value. Because the microprocessor is a digital device, there must be some feature in the DDC panel to convert the digital signal to an output signal which the controlled device can use. Pneumatic actuators can be used to position the controlled device. If this is the case, there must be a component or translator incorporated to provide a digital-to-pneumatic conversion. This is done with a digital-to-voltage converter and voltage-to-pressure converter (aka electric-to-pneumatic transducer). It is the development of these transducers and the development of the computer hardware and software that have made DDC systems cost effective. If the measured value is less than or greater than the desired value the computer circuitry outputs a series of digital impulses that are converted to a modulating signal to the actuator by way of a transducer (electrical-to-pneumatic or electrical-to-electrical). The transducer maintains the computer output signal until readjusted by the computer. Other DDC systems may change the control signal by a series of on-off or open-closed signals to bleed air out of, or put air into the actuator. In all cases there is some interfacing signal device required to isolate the computer output circuitry from the control signal circuitry.

The DDC system can utilize many forms of logic to control the output from a given input. The input signal can by modified considerably by various logic statements as desired, thereby providing a great amount of flexibility in establishing the sequence of operation. With a practical understanding of the HVACR system, the system operator is able to fine tune the control system to provide the most efficient operation possible.

An example DDC controller's output signals (in volts direct current) for both direct acting and reverse acting operation are listed in Table 15-1.

DDC controller's output signals. **Table 15-1**

Temperature (F)	D/A Output (vdc)	R/A Output (vdc)
72	6	9
75	7. 5	7. 5
78	9	6

Because all the setpoints are now programmed within the microprocessor of the DDC system, the owner, energy manager, or system operator has direct control over the environment within the building by dictating the temperature, pressure and humidity setpoints. This prevents the occupants from constantly adjusting the setpoints of a wall thermostat up and down to their individual wishes which causes significant energy waste. An environmental control system can now be provided that is more attuned to the needs of the majority of the occupants and not the individual desires of a select few controlling the room thermostats.

Precise Control

The DDC systems provide the ability to control the setpoint much more accurately than traditional pneumatic systems. One of the inherent flaws of a pneumatic system is that it cannot provide a precise and repeatable setpoint. Pneumatic systems are only modulating control as shown in Figure 15-1. There is always an offset from the setpoint under minimum and maximum load conditions. The DDC system, because it can be programmed to provide proportional, integral, and derivative (PID) control, can provide absolute control of the setpoint under all load conditions. Therefore, if the setpoint is 72°F, it will maintain that setpoint regardless of the load on the HVACR system. This provides considerable energy savings because the controlled variable (temperature, pressure or relative humidity) can be precisely maintained. The digital computer can be programmed to maintain the control point (the actual temperature, pressure or humidity which the controller is sensing) equal to the setpoint using proportional (modulating) control, and adding integral (reset) control. Derivative (rate) control is added for some control sequence (PID). A floating point (moving the controlled device only when the controlled variable reaches an upper or lower limit) may be added as well.

Deadband and Control Sequence

Based on the response time of any particular controlled device a small deaband (above and below the setpoint) can be established to maintain stability. These deadbands, as well as rate of change of the signal to the actuator and minimum length of time between control signal changes, are individually changeable by an authorized operator. The control sequence can be modified by changing the program algorithms, usually without any change in hard ware. The ease of making the changes varies with the system design.

Schedule Changes

Direct digital control and energy management systems provide easy changing of schedule and therefore can reduce the energy waste caused by being on the wrong time and

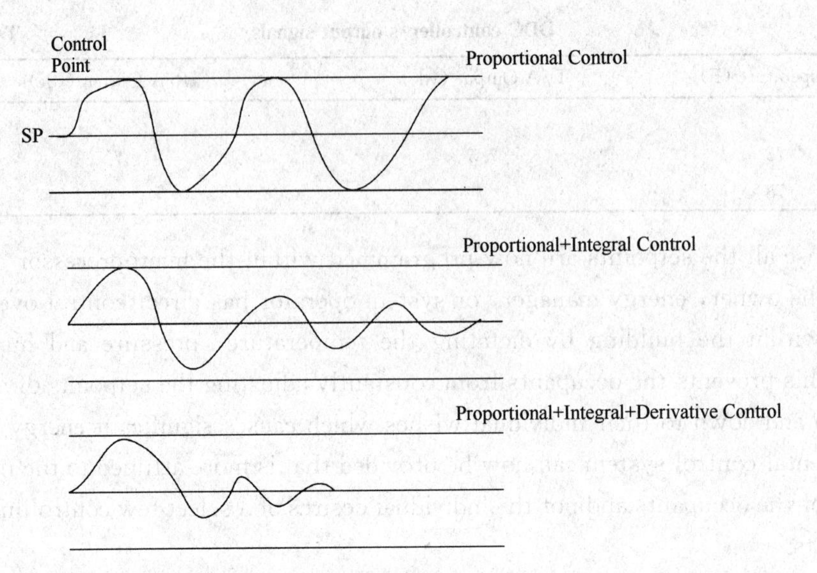

Figure 15-1 Proportional (P, modulating), Proportional and Integral (PI, modulating and reset. It has less drift each cycle.), Proportional, Integral and Derivative (PID, It controls modulation, reset and rate, i. e., less drift each cycle and faster response to get to setpoint.). SP: Setpoint, Drift or offset: The distance from setpoint to control point.

HVACR operation schedule. Day-night schedule, monthly schedules, seasonal schedules, winter-summer schedules, yearly schedules, holiday schedules, etc., can all be quickly changed with simple keyboard entries. For example, in a given facility to change the time clocks from Standard Time to Daylight Saving Time or vice versa might previously have literally required days or longer, whereas with DDC a knowledgeable operator can change the schedule in a few minutes.

Flexibility

A direct digital control system provides greater flexibility in determining how the control loop is to function. The owner-operator-manager has access to software programs which change settings as desired. The system operator can now optimize the control system and provide the most economical operation under all conditions. This is especially important in continually changing conditions within the building or conditioned space such as: number of people, schedule changes, work load changes, and environment changes (lighting, computer, and other heat-generating equipment).

15.2 Energy Management Control Systems

Most DDC systems, in addition to providing local loop control, provide energy management functions that are usually associated with supervisory type energy management systems. Historically, energy management systems (EMS) also now known as Building Automated Systems (BAS), or Building Management Systems (BMS) and various other names, were installed separate form, and in addition to, the local loop controls to provide these functions. This results in a local pneumatic control system which was interfaced to a computer system to provide the energy management functions. Now a DDC system can provide both these functions in one system. The energy management functions provided by these systems include cooling demand control, hot and cold deck reset, chilled water reset, deadband, duty cycling, optimized start, stop, etc.

Cooling Demand Control

The DDC system can automatically reduce the fan speed and/or increase the cooling temperature to unload the refrigeration compressor(s). This provides a percentage of load reduction as opposed to the simple on/off function.

Hot and Cold Deck Reset

Because the local loop control is being set by the DDC system, the hot/cold deck temperatures can be controlled directly, which allows the hot and cold valves to be positioned independently of each other. The heating valve can be commanded, for example, to be closed during the cooling season so there ison overlap of the heating and cooling function.

Chilled Water Reset

By directly controlling the capacity of the chiller, the water temperature can be set at any value desired. This can be a function of outside air temperature, building load, or a combination of both. It assures the most efficient operation of the chiller no matter what the load or outside temperature may be.

Deadband

By direct control of the setpoints of the various systems, a deadband can be programmed into the control algorithm to provide a separation of the heating and cooling setpoints, i. e. , the heating setpoint may be set at 70F and the cooling set for 75F. Between these two temperatures, the system keeps the heating valve and cooling valve closed.

15. 3 Control Components

The typical components of an electric or electronic system are: sensors, control wiring paths, controllers, controlled devices and actuators, relays and switches.

Sensor

A sensing element, either internal or remote of the controller, measures the controlled variable (temperature, humidity and pressure) and sends a signal back to the controller.

A temperature sensor senses a change in temperature. There are two general types of electronic temperature sensing elements, thermistor and thermocouple. A thermistor or resistance temperature detector (RTD) senses a change in temperature with a change in electrical resistance. A thermocouple senses a change in temperature with a change in voltage.

A humidity sensor (hygrometer) is used to measure relative humidity or dew point. An electronic hygrometer senses changes in humidity from either change in capacitance or resistance in the electronic circuitry.

An electronic pressure sensor senses changes in pressures using mechanical devices and then converts this signal to produce current or voltage.

Control Wiring

The control wiring conveys the electricity to the various controllers. For electrical control systems the voltage is 24 volts ac (low voltage), or 110/220 volts ac (line voltage). Electronic control systems use a voltage of 0 to 15 volts dc, or a current of 4 to 20 milliamps dc.

Controller

A controller is a device designed to control a controlled device such as an air damper or a water valve to maintain temperature (thermostat), humidity (humidistat), or pressure (pressurestat). A controller, shown in Figure 15-2, may be direct acting (D/A) or reverse acting (R/A). A direct acting controller increases its outgoing/output signal (electric) or branch line pressure (pneumatic) as the condition it is sensing increases. It decreases its signal as the condition it is sensing decreases. A reverse acting controller decreases its signal as the condition it is sensing increases. It decreases its signal as the condition it is sensing increases.

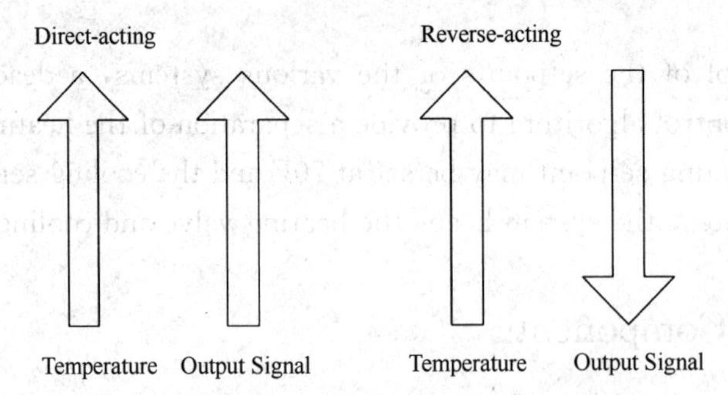

Figure 15-2 Direct-acting and Reverse-acting Controllers.
Temperature is the controlled variable.

Some other important terms you'll need to know are throttling range and setpoint. Throttling range (TR) is the change in the controller condition necessary for the controller output to change over a certain range. For example, if a DDC controller (a thermostat in this example, either direct acting or reverse acting) has a 6 degree throttling range it means the electrical signal output will vary from 6 to 9 volts direct current (vdc) over a 6 degree change in temperature. Another system is pneumatic. One of its controllers (a thermostat, either direct acting or reverse acting) has a 4 degree throttling range. In this example the thermostat's branch line output will vary from 3 to 15 psig over a 4 degree change in temperature.

Setpoint (SP) is the point at which the controller is set and the degree of temperature, or percent relative humidity, or pressure to be maintained.

A thermostat (T-stat) may be either direct or reverse acting. The standard thermostat has a temperature range of 55°F to 85°F. Generally, the throttling range is between 2 and 12 degrees. As an example, a direct acting thermostat with asetpoint of 72°F and a 4 degree throttling range (70°F to 74°F). A direct acting thermostat means that a rise in space temperature causes a rise in the output of the thermostat. Therefore, when the room temperature is at or below 70°F the thermostat signal output is at its minimum. As the space temperature goes above 70°F the output signal will continue to increase until the setpoint temperature is reached or the controller output reaches maximum. Another example would be a room that has a reverse acting thermostat with a 6 degree throttling range and a setpoint of 72°F. The control sequence is: at 75°F, the signal output would be minimum (a rise in space temperature is a decrease in output) and at 69°F the signal output would be maximum.

A humidistat measures humidity in the air and may be either direct acting or reverse acting. For example, reverse acting humidistat controls a normally closed two-way steam valve. As the relative humidity drops in a condition space the output signal to the valve is increased, opening the valve and allowing steam to enter the humidifier.

A pressurestat senses pressure and sends a signal to a controlled device. A static pressure sensor (SPS) in a VAV system is an example. As the static pressure increases in the duct where the SPS is located it senses the increase in pressure and sends a signal to the VFD to slow the fan speed. Pressurestats are also used to make or break and electrical, electronic or pneumatic control signal.

Electric controllers can control flow using the following methods: proportional (modulating), two-position timed, or floating control. They may be single-pole, double-throw (SPDT), or single-pole, single-throw (SPST). Proportional control uses a reversible motor with a feedback potentiometer. Two-position control is used simply to start or stop a device or control a spring-return motor. Two-position timed control uses SPDT circuits to actuate unidirectional motors. Floating control uses a SPDT circuit with a reversible motor.

An electronic direct digital controller (DDC) uses digital computers to receive electronic signals from sensors and converts the signals to numbers. The digital computer (microprocessor or microcomputer) compares the numbers to design conditions. Based on this comparison, the controller then sends out an electronic or pneumatic signal to the actuator.

Direct Digital Control differs from pneumatic or electric- electronic control in that the controller's algorithm (sequence of operation) is stored as a set of program instructions in a software memory bank. The controller itself calculates the proper control signal digitally, rather than using an analog circuit or mechanical change. Interface hardware allows the digital computer to receive input signals from sensors. The computer then takes the input data, and in conjunction with the stored algorithms calculates the changes required. It then sends output signals to relays or actuators to position the controlled devices.

Direct digital controllers are classified as preprogrammed or operator-programmable control. Preprogrammed control restricts the number of parameters, setpoints and limits that can be changed by the operator. Operator-programmable control allows the algorithms to be changed by the operator. Either hand-held or console type terminals allow the operator to communicate with and, where applicable, change the controller's programming.

Controlled Device and Actuator

A controlled device is the final component in the control system. It may be a damper for air control or a valve for water or steam control. Attached to the damper or valve is an actuator. The actuator receives the signal from the controller and positions the controlled device. Actuators are also known as motors or operators.

Some important term to understanding the workings of controlled devices are: normally open, normally closed and actuator spring range. The terms normally closed (NC) and normally open (NO) refer to the position of a controlled device when the power source, compressed air in a pneumatic system, or electricity in an electrical, electronic or DDC system, is removed. A controlled device that moves toward the closed position as the output pressure from the controller decreases is normally closed. A controlled device that moves toward the open position as the output pressure decreases is normally open. The spring range of an actuator restricts the movement of the controlled device within set limits. Automatic dampers used in HVACR systems may be either single blade or multiple blade. Multiple blade dampers are either parallel blade or opposed blade. Parallel blade dampers are diverting and opposed blade dampers are non-diverting. Parallel blade dampers are typically used in the return air duct and in the outside air duct where the two ducts run parallel to each other into the mixed air plenum. The dampers are placed in the ducts so that when they open the return air and the outside air is directed into each other in a mixing application. Opposed blade dampers are used for mixing and volume control applications. The dampers may be installed either normally open or normally closed.

Damper actuators position dampers according to the signal from the controller. A pneumatic systems compressed air pressure may operate the actuator in either a two-position or proportioning manner. Inside the actuator the control air pressure expands the diaphragm around the piston and forces the piston outward against the spring, driving the pushrod out. As air pressure is increased the pushrod is forced to the maximum of the spring range. As air is removed from the actuator, the spring's tension drives the piston towards its normal position. For example, a damper with a proportioning actuator has a spring range of 3 to 7 psig. The actuator is in its normal position when the air pressure is 3 psig or less. Between 3 and 7 psig the stroke of the pushrod is proportional to the air pressure, for instance, 5 psig would mean that the pushrod is half-way extended. At or above 7 psig the maximum stroke is achieved.

Dampers can be sequenced by selecting actuators with different spring ranges. For example, two normally closed dampers operating from the same controller control the air to a conditioned space. The top damper operates from 3 to 7 psig. The bottom damper operates between 8 and 13 psig. Bothe dampers are closed at 3 psig. At 7 psig the top damper is full open and the bottom damper is closed. At 8 psig the top damper is full open and the bottom damper is starting to open. At 13 psig both dampers are full open. Damper actuators may be directly or remotely connected to the damper. The damper position, normally open or normally closed, is determined by the way the damper is connected to the actuator. In other words, if the damper closes when the actuator is at minimum stroke, the damper is normally closed. If, on the other hand, the damper opens when the actuator is at minimum stroke, the damper will be normally open.

Control valves are classified according to (1) their flow characteristics, such as quick opening, linear or equal percentage; (2) their control action, such as normally open or normally closed; and (3) the design of the valve body such as two-way, three-way, single seated or double seated. Flow characteristic refers to the relationship between the length of the valve stem travel expressed as a percent and the flow through the valve expressed as percent of full flow. For example, the quick opening valve has a flat plug, which gives maximum flow as soon as the stem starts up. This type of valve might deliver 90% of the flow when it's open only 10%. Therefore, a typical application for a quick opening valve might be on a stream preheat coil where it is important to have a lot of fluid flow as quickly as possible. By comparison, in a linear valve the percent of stem travel and percent of flow are proportional. For instance, if the stem travel is 50% the flow is 50%. In the equal percentage valve each equal increment of stem travel increases the flow by an equal percentage. For example, for each 10% of stem travel for a particular equal percentage valve the flow is increased by 50%. In other words, at 30% open the flow is 8%, when the stem travels to 40% open the flow is 12% (8% + 4%) and at 50% travel the flow is 18% (12% + 6%), etc. At 90% stem travel the flow is 91.125%.

There are two general categories of electric actuators. One is the solenoid type of actuator. It consists of a magnetic coil operating, a moveable plunger. It is limited to the operation of smaller controlled devices. Most solenoid actuators are two-position. The other category of actuators is known as a motor. Motors are further classified as unidirectional, spring-return or reversible.

EXAMPLE 15-1 Control Subsystems of Four-Pipe Heating and Cooling System

Figure 15-3 shows an air handling unit one of several in this pneumatically controlled four-pipe (two supply pipes and two return pipes) heating and cooling system. The heating and cooling coils have a two-way variable volume control valve in the return pipe. The conditioned space thermostat is direct-acting (D/A). The heating valve is normally open (NO) operating from three to eight pounds (psig). The cooling valve is normally closed (NC) operating from 10 pounds to 13 pounds (psig).

As the conditioned space thermostat (T) senses a temperature above setpoint it sends an increasing signal to the valves. At 3 pounds the heating valve starts to close and continues to close is on the thermostat senses a temperature above setpoint. At 8 psig the heating valve is full closed. If the thermostat is still sensing a room temperature above setpoint the pneumatic air pressure continues to rise. There is a 2 psig deadband between the heating valve and the cooling valve when no hot water is going to the heating coil and no chilled water is going to the cooling coil. As room temperature continues to rise the cooling valve receives a pressure of 10 pounds and starts to open. It will continue to open if the thermostat is not satisfied until it reaches 13 pounds. At this time the cooling valve is full open.

When the space starts to cool below thermometer setpoint the pneumatic air pressure begins to drop from 13 pounds towards 10 pounds. At 10 pounds the chilled water valve is full closed. There is no chilled water or hot water going to their respective valves between 10 and 8 pounds. If the space is still too cold the pressure continues to drop and at 8 pounds the heating valve starts to open and is full open at 3 pounds and receiving full flow of hot water.

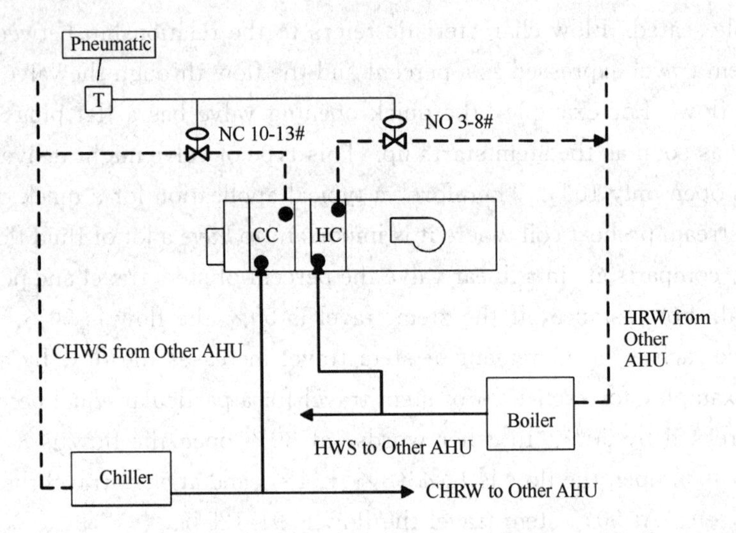

Figure 15-3 Four-pipe heating and cooling system. Direct acting thermostat. Deadband is between 8 and 10 psig.

EXAMPLE 15-2 Control Subsystems of Multizone with Reheat System

Figure 15-4 is a multizone system with central cooling and conditioned space reheat. There is a thermostat for each room which makes it multizone. This central cooling has a three-way valve and a sensor in the supply discharge duct. The "A" port is water coming from the coil. The "B" port is the water coming through the bypass. The "AB" port is water going to the chiller. The "A" port is NC and "B" port is NO. The discharge duct temperature controller is direct-acting (D/A). With a rise in air temperature in the discharge duct an increasing signal will be sent to the chilled water valve to close the bypass "B" port and open the "A" port sending new chilled water through the coil.

The cool air will be discharged into each room (or zone). Let's say the rooms are numbered 1, 2 and 3 left to right. Rooms 1 and 2 need only cooled air. Room 3, however, is overcooled. The direct-acting thermostat senses the drop in temperature in that room and sends a decreasing signal to the three-way valve in the heating pipe to open the NO "A" port and close the NC "B" port. The system would fail to the heating mode. Hot or warm water flows through the reheat coil heating the supply air into the space. These systems can be made variable volume by a designing or retrofitting to two-way valves.

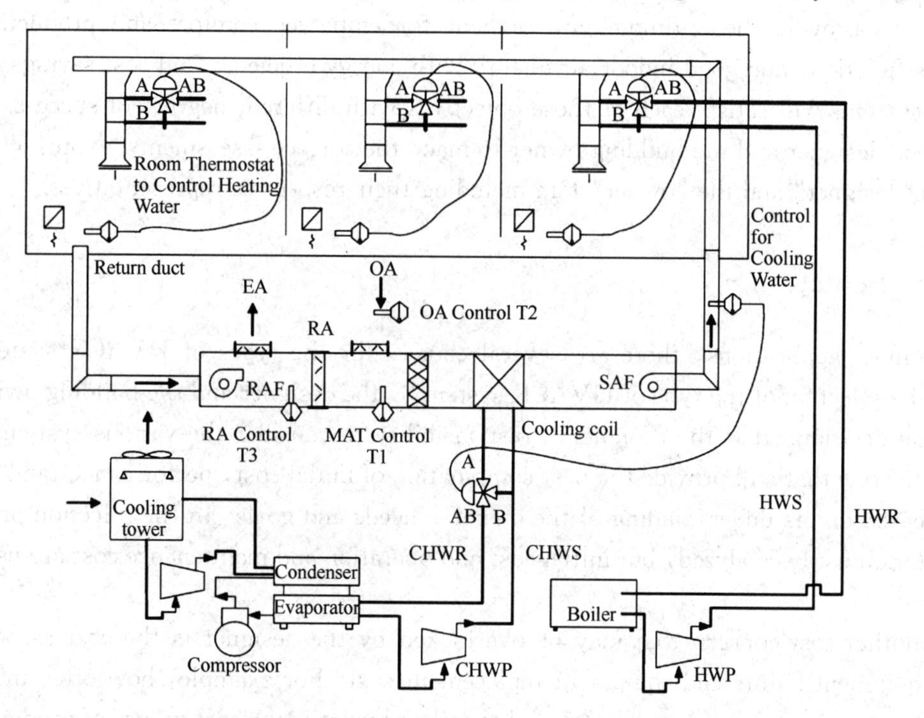

Figure 15-4 Multizone system with central cooling and conditioned space reheat.

Unit 16 System Selection and Optimization

The purpose of an HVACR system is to provide and maintain environmental conditions within an area called the "conditioned space". The type of system selected is determined by the mechanical designer's knowledge of systems and the building owner's financial and functional goals. The commercial system selected for a particular application endeavors to provide the optimum environment for employee comfort and productivity, process function, and good indoor air quality with energy efficiency and cost savings. Different systems will satisfy each of these objectives with different degrees of success. It is up to the designer and the building owner to make the correct assessments. Note: When I say the"designer" and the "owner" I'm including their respective representatives.

16.1 Selection

In most applications, there are several choices for the types of HVACR system to use. The selection of the type of HVACR system by the designer and the building owner is a critical decision. It is the designer's responsibility to consider the various systems and select the one that will provide the best combination of initial cost, performance, and reliability based on his understanding of the owner's needs and goals. In the selection process all factors must be analyzed, but initial cost and operation and maintenance cost are usually foremost.

Another cost concern that may be overlooked by the designer is the cost associated with equipment failure and equipment replacement cost. For example, how often might a selected system or component be expected to fail and what is the cost in loss of product and production? How long will the system be down? How will the comfort, safety and productivity of the occupants be affected by such a failure and what is the cost?

Depending on the owner's goals each of these concerns has a different priority. Most owners may not have detailed knowledge or understanding of the advantages of the different types of systems. Therefore, it is normally up to the designer to make the equipment selection. On the other hand, the designer may not have a complete understanding of the owner's financial and functional goals. For these reasons the best situation is when both the designer and owner are involved in the HVACR selection process.

The first step in the selection process is for the designer to ascertain and document from the owner the desired environmental conditions for the building or conditioned space. The designer must also learn and document the restrictions placed on the system design. For example, what is the required equipment space for a particular system versus what is available? Unfortunately, it is the nature of the business that very few projects allow as much detailed evaluation of all conditions and alternatives as some would like. Therefore, the designer must also rely on common sense and subjective experience to narrow the choice of systems.

Step two in the selection process is determining the building's heating and cooling loads. For example is the cooling load mostly sensible or latent? Is the load relatively high or low per square foot of conditioned area as compared to other similar buildings? Is the load uniformly distributed throughout the conditioned space? Is it relatively constant or does it vary greatly? How does the load vary with time and operating conditions? Determining the heating and cooling loads establishes the system's capacity requirements. Cooling loads and humidity requirements are used to size air conditioning (comfort and process cooling) system. In other systems, heating or ventilation may be the critical factors in sizing and selection. For example, a building may require a large air handling unit and duct system to provide huge quantities of outside air for ventilation or as make-up air to replace air exhausted from the building. In other buildings, in colder climates for instance, heating may be the determining factor on equipment size. The physical size of the equipment can be estimated from the heating and cooling load information alone. This information can help in the choice of systems to those that will fit the space available.

There are also choices to be made depending on whether the system is to be installed in a new building or an existing building. In existing buildings, for example, the HVACR system was designed for the loads when the building was built. This means if new systems are to be integrated with existing ones (in order to keep costs down or for other reasons) the new or retrofitted systems must be adaptable to existing equipment, ductwork and piping, and new equipment or systems must fit into existing spaces. If new systems are to perform properly when tied in with existing systems the old and the new must be looked at carefully and in its entirety. The designer will need to determine how a change to one part of a system will affect another part and how a change in one system will affect another system. The number of choices is narrowed further to those systems that will work well on projects of a given application and size and are compatible with the building architecture.

16.2 Selection Guidelines

Each of the following issues should be taken into consideration each time an HVACR system is selected.

Financial Factors

Initial cost

Operating cost

Maintenance and repair cost

Equipment replacement or upgrading cost Equipment failure cost

Return on investment (ROI)

Building Conditions

New or existing building or space Location

Orientation

Architecture

Climate and shading

Configuration

Construction

Codes and standards

Building Use

Occupancy

Process equipment

Energy

Types

Availability

Reliability

Control Scheme

Zone control

Individual control

The basic types of HVACR systems used in commercial buildings are build-up, unitary and split. A built-up heating, ventilating and air conditioning system is one that is custom designed for the building in which it is installed. It typically consists of a central plant with equipment and components that generate and distribute heated or cooled air or water energy to the conditioned spaces within the building. A unitary system is an air conditioning unit that provides all or part of the air conditioning functions (heating, cooling, fan, filter, compressor, condenser, controls) in one or a few assemblies, in essence in one package. Types of unitary systems include package units, rooftop units, window-mounted air conditioners and heat pumps, through-the-wall air conditioners and heat pumps, and packaged terminal air conditioners and heat pumps. A split system has an indoor section and matching outdoor section connected by refrigerant tubing. The indoor section typically consists of a fan, an indoor cooling coil, a heating function, and filters while the outdoor section houses the compressor and condenser. The unitary system is all-in-one and the split system is split into two sections, an outdoor section and an indoor section. Systems are further classified as air and water, all-air, or all-water. Water systems are also called hydronic systems.

Unitary System

Unitary systems are selected when it is decided that a central HVACR system is too large or too expensive for a particular project or a combination system (central and unitary) is needed for certain areas or zones to supplement the central system. For example, unitary systems are frequently used for perimeter spaces in combination with a central all-air system that serves interior building spaces. This combination will usually provide greater temperature and humidity control, air quality, and air (conditioned air and ventilation air) distribution patterns, than is possible with central or unitary units alone. As with any HVACR system both the advantages and the disadvantages of unitary systems should be carefully examined to ensure that the system selected will perform as intended for the particular application.

A solid understanding of the various types of commercial HVACR systems and their selection is important if you are the energy manager or facilities engineer. This position often calls for being the owner's representative, working with others to ensure that the owner gets the environmental system that will best fit his needs. Following are some of the advantages and disadvantages to consider when selecting unitary system.

Temperature Control and Airflow

Individual room control (on/off and temperature) is simple and inexpensive. However, because temperature control is usually two-position there can be swings in room temperature. Also, the room occupant has limited adjustment on air distribution and airflow quantity which are fixed by design. Ventilation airflow quantity is also fixed by design as are the sizes of the cooling and condenser coils. On the plus side, ventilation air is provided whenever the unit operates.

Humidity Control

Unitary systems can provide heating and cooling capability at all times independent of other spaces in the building but basic systems do not provide close humidity control. However, close humidity control is not needed for most applications. But, if needed, in computer room applications or the like, close humidity control can be accomplished by selecting special purpose packaged units.

Manufacturing

Manufacturer-matched components have certified ratings and performance data and factory assembly provides improved quality control and reliability. There are a number of manufacturers so units are readily available but equipment life may be short (10-15 years) as compared to larger equipment which may have life expectancies of 20-25 years. Manufacturers' instructions and multiple-unit arrangements simplify the installation through repetition of tasks. However, from an architectural or esthetic point of view appearance may be unappealing.

Maintenance and Operation

An advantage of unitary systems is only the one unit and one temperature zone is affected if a unit malfunctions. Also, less mechanical and electrical space is required than with central systems. And, in general, trained operators are not required as might be with more sophisticated central systems. However, maintaining the units is more difficult because of the many pieces of equipment and their location such as on roof tops or in occupied spaces. Also, condensate can be a problem if proper removal or drainage is not provided.

Other disadvantages are that air filtration options may be limited and the operating sound levels can be high.

Cost and Energy Efficiency

Initial cost is usually low but operating cost may be higher than for central systems. This will be the case when the unitary equipment efficiency is less than that of the central system components. Also, energy use may be greater because fixed unit size increments require over-sizing for some applications and outdoor air economizers are not always available to provide low cost cooling. However, for leased space applications such as offices, retail space, R&D labs, computer rooms, etc., energy use can be metered directly to each tenant. Also, units can be installed to condition just one space at a time as a building is completed, remodeled, or as individual areas are leased and occupied. Another energy management opportunity with unitary systems is that units serving unoccupied spaces can be turned off locally or from a central point without affecting occupied spaces.

Split System

The guidelines for selection of split and variable refrigerant flow (VRF) systems are similar to the information provided above for unitary systems. However, two other concerns that need to be addressed are length of refrigeration piping and ventilation air. Length of piping is not the concern it once was as newer systems have steadily increased maximum refrigerant pipe length, in some cases, to many hundreds of feet, however, meeting ANSI/ASHRAE ventilation code requirements is still a concern as split systems and VRF systems do not inherently provide ventilation so a separate ventilation system may be necessary.

16.3 Optimization

In the 1950s, air conditioning systems dramatically changed the way we live in the United States. As HVACR systems became more reliable, efficient and controllable, we were no longer dependent on the weather for work or leisure. We made the environment adapt to our needs. In fact, we started cooling to temperatures lower than the temperatures to which we had previously heated. Today, indoor climate control has become so reliable and affordable it is common in industry and homes alike as almost all commercial buildings have HVACR systems and many U.S. households have air conditioning. The goal of HVACR systems continues to be to provide a high degree of occupancy comfort and indoor air quality and to maintain environmental conditions for work processes while holding operating costs to a minimum.

Evaluate System Efficiency

There are many ways to evaluate whether or not the system is efficient. For instance, general appearance, equipment down-time, maintenance records, maintenance costs and occupant complaints are items that need to be taken into consideration. However, one of the best ways to evaluate how the system is performing, if energy usage costs are a major concern, is by comparing energy used over several years. To make this comparison, develop a Building Energy Use Number (BEUN), as listed in Table 16-1.

Example BEUN. LPG: liquid petroleum gas, kWh: kilowatt hour, mcf: 1000 cubic feet, gal: gallon, mlb: 1000 pounds, * Consult energy supplier for conversion factor. **Table 16-1**

Commodity	Unit	Units Consumed Per Year	Btu Conversion	Btu/year
Electricity	kWh	1000	3,413 Btu/kWh	3,413,000
Natural Gas	therm	10	100,000 Btu /Therm	1,000,000
Natural Gas	mcf		*	
LPG	gal		*	
Fuel Oil	gal		*	
Coal	ton		*	
Steam	mlb		Steam table	
Other				
Total Btu				4,413,000
Building sf				10,000
Btu /sf/year				441.3

To develop the BEUN gather the building's utility records for the past 24 months. Obtain the rates charged by energy suppliers including commodity rates, demand rates, discounts, taxes, on- and off-peak rates, power factor rates, ratchet charges, etc. Determine the total energy used by the building's HVACR systems for one year. This will include electricity, natural gas and oil and any other commodity (such as liquid petroleum gas) used to operate the HVACR system. Convert the energy used to Btu per year. Divide the Btu per year by the square feet of the building's conditioned space. This is the BEUN for the base year in Btu per year per square foot.

Compare the BEUN with other similarly constructed and used buildings in the area or across the country. Information on BEUN for similar buildings can be obtained from Federal, state or local government agencies or the utility company. Additional useful information for evaluating the HVACR system and energy efficiency would include occupancy data (hours the building is occupied, type of work performed and number of people per shift), weather data for the base year and present year, data on how the building is constructed, and HVACR operation and maintenance logs and manuals.

Establish Efficiency Goals

Organize a team consisting of staff, maintenance personnel, consultants, contractors and energy suppliers. Gather all information needed to develop the BEUN and evaluate HVACR systems. Identify type of HVACR systems and their interaction with each other.

Determine system performance. Evaluate system maintenance. State system problems and opportunities and set efficiency goals. Determine what resources will be needed to achieve goals, including staffing and money. Assign responsibilities for achieving goals and set a time schedule to reach goals. Monitor results. Review and revise goals as needed.

Energy Systems

It's generally assumed that the HVACR systems and lighting systems account for most of a building's energy use. HVACR energy consumption is affected in part by the common practice of specifying oversized heating and cooling equipment to compensate for the energy inefficiency in a building's design and construction. The following are energy conservation opportunities (ECO) for HVACR systems and subsystems.

HVACR System ECO

- Compare field measurements (air, water, steam and electrical) with the air or water balance report, commissioning report, and fan, pump, and motor curves to determine if the correct amount of air and water is flowing.
- Use nameplate data to prepare an up-to-date list of motors for fans, compressors, pumps, etc., and list routine maintenance to be performed on each.
- Routinely check time clocks and other control equipment for proper operation, correct time and day, and proper programming of on-off setpoints.
- Reduce or turn off heating and cooling systems during the last hour of occupancy.
- Close interior blinds and shades to reduce night heat loss in the winter or night and solar heat gain in the summer or day. Repair or replace damaged or missing shading devices.
- Inspect room supply air outlets and return and exhaust air inlets, diffusers, grilles and registers.
- Clean ducts. Open access doors to check for possible obstructions such as loose insulation in lined ducts, loose turning vanes and closed volume or fire dampers. Adjust, repair or replace these items as necessary.
- Reduce outdoor air intake quantity to the minimum allowed under codes by adjusting outdoor air dampers. Maintain a rate of 15-25 cubic feet per minute (cfm) of air per person. Maintain outside air dampers.
- List automatic and gravity dampers and routinely check that they open and close properly. Adjust linkage or replace dampers if the blades do not close tightly.
- Replace unsatisfactory automatic dampers with higher quality opposed blade or parallel blade dampers with seals at edges and ends to reduce air leaks. Readjust position indicators as needed to accurately show the position of all dampers.
- Regularly clean or replace dirty or ineffective filters.
- Clean coils and other heat exchangers.
- Ensure that all fans rotate in the proper direction.
- Check fan, pump, or compressor motor voltage and current.
- Adjust fan speed, inlet guide vanes, or VFD (variable frequency drive) for proper air-

flow.

- Measure total static pressure across fans and total dynamic head across pumps.
- Maintain correct belt tension on fan-motor drives.
- Check drives for misalignment.
- Discontinue use of unneeded exhaust fans.
- Rewire toilet exhaust fans to only operate when lights are on or there's a signal from an occupancy sensor.
- Check pump suction and discharge pressures and plot differential pressure on the pump curve.
- Close the discharge valve if the pump circulation is more than 10 percent greater than required flow.
- Reduce pump impeller size for greater energy savings.
- Adjust pump speed, impeller, or VFD (variable frequency drive) for proper water flow.
- Properly adjust and balance air and water systems.
- Adjust controls.
- Install a time clock or automated energy management system that will reduce heating and cooling.
- Close some air conditioning supply and return ducts for HVACR systems operating in lobbies, corridors, vestibules, public areas, unoccupied areas or little-used areas. Disconnect electrical or natural gas heating units to these areas.

Airside ECO

- Design or retrofit system for lowest pressure needed
- Design or retrofit system for lower airflow
- Install balancing dampers
- Air balance system
- Avoid installing restrictive ductwork on inlets and discharges of fans
- Install variable volume systems where applicable
- Reduce equipment on-time
- Use economizers
- Maintain systems including controls
- Verify fans are rotating in correct direction
- Clean fan blades
- Clean filters and coils
- Repair leaks in duct system
- Insulate duct
- Reduce resistance in system
- When reducing airflow change fan speed instead of closing main dampers

Waterside ECO

- Design or retrofit system to lowest pressure needed

- Design or retrofit system for lower water flows
- Use primary-secondary circuits and variable flow systems where applicable
- Install flow meters and provide balancing valves
- Water balance system
- When reducing water flow change pump speed or trim impeller instead of closing main valves
- Avoid installing restrictive piping on the inlets and discharges of pumps
- Reduce equipment on-time
- Use economizers
- Ensure pumps have correct rotation
- Maintain systems including controls
- Clean water coils, condensers, evaporators and cooling towers
- Clean strainers and restrictive valves
- Repair or replace leaking valves and pipes
- Re-pipe crossed-over piping
- Insulate pipe
- Reduce resistance in system
- Keep debris and other obstructions away from coils and towers
- Use two speed or variable speed fans on cooling towers
- Raise delta T on heat exchangers

Refrigeration Side ECO

- Adjust controls on multiple staging systems and inspect that staging functions properly. For example: second compressor doesn't energize until the first compressor can no longer satisfy the demand, etc., through all stages.
- Clean condenser coils on air-cooled system. Remove debris restricting airflow. Also remove debris or restrictions from evaporator coils. See Table 16-2.
- Defrost evaporator coils if iced. Determine the cause of icing and correct it (normally low air volume or low refrigerant charge)
- Clean scale build-up in water-cooled condensers.
- Record normal operating temperatures and pressures and check gauges frequently to ensure conditions are met.
- Check for proper refrigerant charge, superheat, and operation of the metering device.
- Repair leaking compressor valves.
- Repair leaking liquid line solenoid valves and clean liquid line strainers.
- Experiment with chilled water supply temperature while maintaining an acceptable comfort level.
- Increase temperatures to reduce energy used by the compressor or decrease temperature to reduce water pump horsepower.
- To reduce condensing temperatures on air-cooled condensers consider (1) increasing the air volume through the condenser by increasing the fan speed (2) adding more air-

cooled condensers in parallel to increase coil heat transfer surface area or (3) replace the existing condenser coil with one that has a larger surface area.

Table shows increase in brake horsepower per ton of cooling with restriction (inside or outside) of heat exchangers (condenser and/or evaporator). Table 16-2

System Operating Condition	Evaporator Temperature	Condenser Temperature	Tons of Cooling	Bhp	Bhp per ton	Increased Bhp per ton
Normal Operation	45	105	17	16	0.93	
Restricted Condenser	45	115	15.6	18	1.12	20%
Restricted Evaporator	35	105	13.8	15	1.10	18%
Both Restricted	35	115	12.7	16	1.29	39%

Boiler

- Ensure the proper amount of air for construction is available. Check that primary and secondary air can enter the boiler's combustion chamber in regulated quantities and at the correct place.
- Inspect boiler gaskets, refractory, brickwork and castings for hot spots and air leaks. Defective gaskets, cracked brickwork and broken casings allow uncontrolled and varying amounts of air to enter the boiler and prevent accurate fuel-air ratio adjustment.
- Perform a flue-gas analysis. Take stack temperatures and oxygen readings routinely and inspect the boiler for leaks.
- Repair all defects before resetting the fuel-air ratio. Consider installing an oxygen analyzer with automatic trim for larger boilers. This device continuously analyzes the fuel-air ratio and automatically adjusts it to meet the changing stack draft and load conditions.
- Check that controls are turning off boilers and pumps as outlined in the sequence of operations. Observe the fire when the boiler shuts down. If it does not cut off immediately check for a faulty solenoid valve and repair or replace it as needed.
- Adjust controls on multiple systems so a second boiler will not fire until the first boiler can no longer satisfy the demand. Make sure that reset controls work properly to schedule heating water temperature according to the outside air temperature.
- Experiment with hot water temperature reduction until reaching an acceptable comfort level.
- Install automatic blowdown controls. Pipe the blowdown water through a heat exchanger to recover and reuse waste heat.
- Inspect boiler nozzles for wear, dirt or incorrect spray angles. Clean fouled oil nozzles and dirty gas parts.
- Replace all oversized or undersized nozzles. Adjust nozzles as needed.
- Verify that fuel oil flows freely and oil pressure is correct. Watch for burner short-cycling.
- Inspect boiler and pipes for broken or missing insulation and repair or replace is as nee-

ded.

- Clean the fire side and maintain if free from soot or other deposits.
- Clean water side and maintain if free from scale deposits.
- Maintain the correct water treatment. Remove scale deposits and accumulation of sediment by scraping and/or treating chemically.

Heat Recovery

The objective of heat recovery systems is to reduce the energy consumption and cost of operating a building by transferring heat between two fluids. Exhaust air and outside air is one example. In many cases, the proper application of heat recovery systems can result in reduced energy consumption and lower energy bills while adding little or no additional cost to building maintenance or operations. However, if it cannot be shown that the benefits of a heat recovery system outweigh the cost building owners will not be motivated to make a financial investment in such a system.

During the past 50 years building owners and other commercial energy end-users have had to find ways to cope with increasing uncertainty about the supply and economic volatility of nuclear, electrical and fossil fuels used to generate energy for their facilities. Indeed, weather, politics, and market forces play a significant role in determining the availability of energy and its cost. End-users need only to recall the power shortages in the last decade that plagued sections of the U. S. crimping supplies and sending energy prices soaring. Those with a greater sense of history are aware of the oil shortages of the mid-1970s. I was just starting in HVACR then and I remember that first oil embargo and the threat that gasoline could go as high as sixty cents ($0. 60) per gallon.

The continuing unrest suggests that an oil crisis is not only possible but probable. It is just another wakeup call that energy created by fossil fuels will not always be readily accessible. It also serves notice about the need to reduce our reliance on such energy sources to better insulate business from forces beyond our control. Designers and facility managers of commercial buildings need to focus their efforts on energy conservation as well as maximizing occupant comfort and process function. While many of the conservation measures implemented are voluntary regulators will also continue to mandate energy conservation strategies.

One energy conservation measurement worth considering is heat recovery which captures waste heat from one fluid and transfers it to another cooler fluid to be used in a heating application. An example of a heat recovery system is heated exhaust air to wintertime outside air coming into an AHU. A heat recovery system reduces energy consumption by eliminating the need to generate new heat and in turn lowers building operating costs. Another example is capturing waste heat from the flue gas of a boiler and the reusing that heat to preheat the boiler input water and therefore the amount of heat required to generate steam or hot water is reduced. A third example is capturing waste heat from large ovens or other similar heat-generating equipment to be used for comfort heating.

Additionally, heat recovery systems (also known as heat energy or energy recovery systems) can be used to provide reserve heat (energy) capacity. The reason for having reserve capacity is that many times the implemented energy conservation measures substantially reduce the capacities of HVACR equipment. Therefore, the installed equipment capacities now closely match the design load and less reserve capacity is available for new projects that may substantially change the building HVACR system. For example, many installed systems may not have enough reserve capacity to make needed changes to accommodate increased outdoor air requirements in order to satisfy indoor air quality (IAQ) concerns. When more outside air is needed a heat recovery system can help to offset the increased energy cost to heat up or cool down the increased volume of outside air.

There are three basic types of heat recovery systems: comfort-to-comfort, process-to-comfort, and process-to-process. The types of heat exchangers for these systems include rotary wheel, fixed plate, heat pipe, and run-around coil. The effectiveness of a heat exchanger (coil, plate heat exchanger, heat wheel, etc.) in a heat recovery system is dependent upon three factors: (1) the temperature difference of the fluids circulated through the exchanger; (2) the thermal conductivity (ability to conduct heat) of the material (copper, aluminum, steel, etc.) in the exchanger; and (3) the flow pattern (e. g., counter flow or parallel flow) of the fluids. Heat transfer is greatest in counter flow exchangers. Counter flow is when "Fluid A" enters on the same side of the exchanger that "Fluid B" is leaving as shown in Figure 16-1. Parallel flow is when "Fluid A" enters on the same side of the exchanger that "Fluid B" is entering.

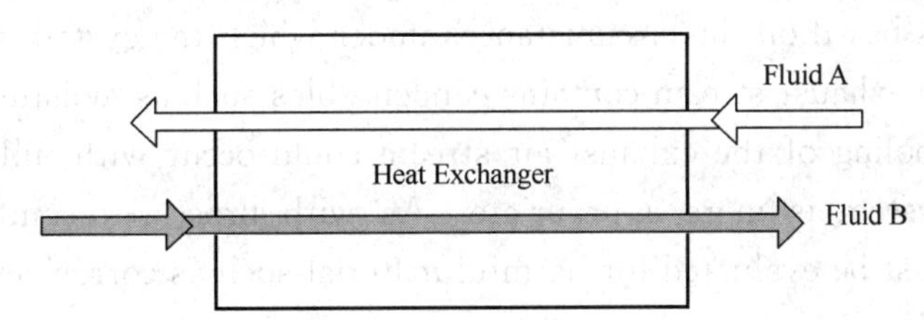

Figure 16-1 Counter flow heat exchanger

Comfort-to-Comfort Heat Recovery System

Comfort-to-comfort heat recovery systems are typically used in HVACR applications. These heat recovery systems capture a building's exhaust air and reuse the energy to precondition the outside air coming into the building. In comfort-comfort applications the energy recovery process is reversible, i. e., the enthalpy (total heat content) of the building supply air is lowered during warm weather and raised during cold weather. Air-to-air heat recovery systems for comfort-to-comfort applications fall into two general categories: sensible heat systems and total heat systems.

Sensible heat recovery systems transfer sensible heat between exhaust air leaving the building and make-up or supply air entering the building. Rotary wheel heat exchangers are used in comfort-to-comfort sensible heat recovery applications. The amount of sensible heat transferred through the air can be determined by

$$\text{Btu}_{hs} = \text{cfm} \times 1.08 \times \text{TD} \times \text{system efficiency} \qquad (16\text{-}1)$$

where Btu_{hs}——sensible heat transfer

cfm——quantity of airflow

1. 08——constant

TD——dry buld temperature difference between the airstreams

For a total heat recovery system, air total heat transfer can be calculated by

$$Btu_{ht} = cfm \times 4.5 \times \Delta h \times \text{system efficiency} \quad (16\text{-}2)$$

where Btu_{ht}——total heat transfer

cfm——quantity of airflow

4. 5——constant

Δh——total heat difference between the airstreams

Process-to-Comfort Heat Recovery System

Process-to-comfort heat recovery systems are generally only sensible heat recovery used during the spring, fall, and winter months. When considering process-to-comfort heat recovery system the process effluent (the exhaust from equipment and the actual work process) must be evaluated for harmful material such as corrosives, condensables (moisture or water vapor), contaminants, and noxious or toxic substances.

Process-to-Process Heat Recovery System

Process-to-process heat recovery systems are typically only sensible heat recovery, usually full recovery, but in some cases, partial recovery can be performed. Determining when to use a process-to-process system for partial sensible heat recovery instead of full sensible recovery is based on the circumstances under which the system will operate. For example, when the exhaust stream contains condensables such as moisture or water vapor and possible overcooling of the exhaust air stream could occur with full recovery then a partial recovery system is more appropriate. As with process-to-comfort systems the process effluent must be evaluated for harmful material such as corrosives, contaminants, and noxious or toxic substances.

Unit 17 International Communication

17.1 How to Write a Scientific Paper

A scientific paper is a written and published report describing original research results. That short definition must be qualified, however, by noting that a scientific paper must be written in a certain way and it must be published in a certain way, as defined by three centuries of developing tradition, editorial practice, scientific ethics, and the interplay of printing and publishing procedures.

1. Title

What is a good title? I define it as the **fewest possible words** that adequately describe the contents of the paper. The title of a paper is a label. It is not a sentence. Because it is not a sentence, with the usual subject, verb, object arrangement, it is really simpler than a sentence (or, at least, usually shorter), but the order of the words becomes even more important.

"Running titles" or "running heads" are printed at the top of each page. Often, the title of the journal' or book is given at the top of left-facing pages and the article or chapter title is given at the top of right-facing pages. Usually, a short version of the title is needed because of space limitations. (The maximum character count is likely to be given in the journal's Instructions to Authors.) It is wise to suggest an appropriate running title on the title page of the manuscript.

Length of the Title

- Not too short

EXAMPLE 17-1 "Studies on Brucella" submitted to the *Journal of Bacteriology*

Analysis Obviously, such a title was not very helpful to the potential reader. Was the study taxonomic, genetic, biochemical, or medical? We would certainly want to know at least that much.

- Not too long

EXAMPLE 17-2 "On the addition to the method of microscopic research by a new way of producing colour-contrast between an object and its background or between definite parts of the object itself" (J. Rheinberg, *J. R. Microsc. Soc.* 1896: 373).

Analysis Much more often, titles are too long. Ironically, long titles are often less meaningful than short ones. A generation or so ago, when science was less specialized, titles tended to be long and nonspecific. That certainly sounds like a poor title; perhaps it would make a good abstract. Without question, most excessively long titles contain "waste" words. Often, these waste words appear right at the start of the title, words such as "Studies on," "Investigations on,"and "Observations on." An opening *A*, *An*, or *The* is also a "waste" word. Certainly, such words are useless for indexing purposes.

Need for Specific Titles

EXAMPLE 17-3 "Action of Antibiotics on Bacteria."

Analysis Is it a good title? In *form* it is; it is short and carries no excess baggage (waste words). Certainly, it would not be improved by changing it to "Preliminary Observations on the Effect of Certain Antibiotics on Various Species of Bacteria." However, most titles that are too short are too short because they include general rather than specific terms. We can safely assume that the study introduced by the above title did not test the effect of all antibiotics on all kinds of bacteria. Therefore, the title is essentially meaningless. If only one or a few antibiotics were studied, they should be individually listed in the title. If only one or a few organisms were tested, they should be individually listed in the title. If the number of antibiotics or organisms was awkwardly large for listing in the title, perhaps a group name could have been substituted. Examples of more acceptable titles are

"Action of Streptomycin on *Mycobacterium tuberculosis*"

"Action of Streptomycin, Neomycin, and Tetracycline on Gram-Positive Bacteria"

"Action ofPolyene Antibiotics on Plant-Pathogenic Bacteria"

"Action of Various Antifungal Antibiotics on *Candida albicans* and *Aspergillus fumigatus*"

Although these titles are more acceptable than the sample, they are not especially good because they are still too general. If the "Action of" can be defined easily, the meaning might be clearer. For example, the first title above might be phrased "Inhibition of Growth of *Mycobacterium tuberculosis* by Streptomycin."

Importance of Syntax

EXAMPLE 17-4 "Mechanism of Suppression of Nontransmissible Pneumonia in Mice Induced by Newcastle Disease Virus" submitted to the *Journal of Bacteriology*

Analysis Unless this author had somehow managed to demonstrate spontaneous generation, it must have been the pneumonia that was induced and not the mice. (The title should have read: "Mechanism of Suppression of Nontransmissible Pneumonia Induced in Mice by Newcastle Disease Virus.")

EXAMPLE 17-5 "Isolation of Antigens from Monkeys Using Complement-Fixation Techniques." submitted to the *Journal of Bacteriology*

Analysis Be careful when you use "using." The word "using" is, I believe, the most common dangling participle in scientific writing. Either there are some more smart dogs, or "using" is misused in this sentence from a manuscript: "Using a fiberoptic bronchoscope, dogs were immunized with sheep red blood cells."

Abbreviations and Jargon

Titles should almost never contain abbreviations, chemical formulas, proprietary (rather than generic) names, jargon, unusual or outdated terminology, and the like.

2. Abstract

An Abstract should be viewed as aminiversion of the paper. The Abstract should provide a brief summary of each of the main sections of the paper: Introduction, Materials and Methods, Results, and Discussion. As Houghton (1975) put it, "An abstract can be defined as a summary of the information in a document."

"A well-prepared abstract enables readers to identify the basic content of a document quickly and accurately, to determine its relevance to their interests, and thus to decide whether they need to read the document in its entirety" (American National Standards Institute, 1979b). The Abstract should not exceed 250 words and should be designed to define clearly what is dealt with in the paper. The Abstract should be typed as a single paragraph. (Some medical journals now run "structured" abstracts consisting of a few brief paragraphs.) Many people will read the Abstract, either in the original journal or in*Biological Abstracts*, *Chemical Abstracts*, or one of the other secondary publications (either in the print editions or in online computer searches).

The Abstract should (1) state the principal objectives and scope of the investigation, (2) describe the methods employed, (3) summarize the results, and (4) state the principal conclusions. The importance of the conclusions is indicated by the fact that they are often given **three times**: once in the Abstract, again in the Introduction, and again (in more detail probably) in the Discussion.

Most orall of the Abstract should be written in the past tense, because it refers to work done.

The Abstract should never give any information or conclusion that is not stated in the paper. References to the literature must not be cited in the Abstract (except in rare instances, such as modification of a previously published method).

The language should be familiar to the potential reader. Omit obscure abbreviations and acronyms. Unless a long term is used several times within an Abstract, do not abbreviate the term. Wait and introduce the appropriate abbreviation at first use in the text (probably in the Introduction). Write the paper before you write the abstract, if at all possible.

Economy of Words

By far the most common fault is the inclusion of extraneous detail in the Abstract. It is of fundamental importance that the Abstract be written **clearly** and **simply.** If you cannot attract the interest of the reviewer in your Abstract, your cause may be lost. Very often, the reviewer may be perilously close to a final judgment of your manuscript after reading the Abstract **alone.** This could be because the reviewer has a short attention span (often the case). However, if by definition the Abstract is simply a very short version of the whole paper, it is only logical that the reviewer will often reach a preliminary conclusion, and that conclusion is likely to be the correct one. Usually, a good Abstract is followed by a good paper; a poor Abstract is a harbinger of woes to come.

When writing the Abstract, examine every word carefully. If you can tell your story in 100 words, do not use 200. Economically and scientifically, it doesn't make sense to waste words. The total communication system can afford only so much verbal abuse. Of more importance to you, the use of clear, significant words will impress the editors and reviewers (not to mention readers), whereas the use of abstruse, verbose constructions is very likely to provoke a check in the "reject" box on the review form.

Useful Expressions and Patterns for Abstract

- *There has been a focus on* energy conservation and recycling in this paper.
- In this paper the author *examines* the possibilities of reducing environmental pollution and airborne pollution.
- The author of this paper *presents* the data *in terms of* basic inputs and outputs for this new type of the computer.
- We *present* some observations, *both theoretical and experimental*, concerning the effects of the planned and market economics.
- This paper *presents results from an investigation* of the chemical composition of the light aromatic hydrocarbons in several samples of petroleum asphalt, *by means of*, a combination of chromatographic, chemical and spectroscopic methods.
- *The brief history of* human development is presented in the first part of the paper.
- *The detailed characterization of* the aspects of language and learning analysis *is presented*.
- The author *describes* the network and technical aspects of a new transmultiplex equipment.
- Performance goals and various approaches to achieve them *are described* in this paper.
- The model *describes* both the static and dynamic aspects of the electric field.
- The results on male and female communication style are *reported*.
- The paper *reports of* the background and history of the brain study project.
- Steps taken to improve the competition in the financial industry *are reported*.
- Figures of merit, program development tools and the required protocol for such an experiment *are introduced* and *evaluated* by means of a sequence of generating functions.
- The software package *accounts for* the contribution of the double spacing or justified or centered text for the word processer.
- The author here *characterizes* the most general expression for the infinite dimensional algebra.
- The paper *deals with* the effects of the global village on the human life and the applied stress of the people in the future world.

- The paper *looks at* what is happening across China as well as at new design techniques and materials now in use.
- This paper *discusses* the intelligence reasons for this variety and *offers an insight into* design and manufacturing methods, plus some cautionary notes on the hazards of misapplication.
- The paper *identifies* the type of data and analyses for the world population.
- The author *identifies* some simple procedures to highlight both hardware and software problems.
- Relationships between the food we eat today and the health of the people *have been stated* hear in the paper.
- This paper *surveys* the informal letter describing tourist activities in an overall viewpoint.
- The author *considers* two specific subjects which may be covered in a computer center audit.
- The psychological process is *analyzed* by using a conventional approach.
- This paper *contains* and analysis of Chinese traditional treatments.
- The scope of the research *covers* both the civil law and the criminal law.
- The theory of the advertisement schemes *is reviewed* in so far as relevant in the present paper.
- This paper *reviews* these applications, summarizes the theory from materials science viewpoint.
- This paper *concentrates on* journalism and the market economy.
- Particular *attention was given* to the evaluation of start-up procedure on reactor behavior.
- *The emphasis is also placed upon* the implications of attitudes taken toward the knowledge economy and the knowledge property right.
- This paper *provides* the quantitative background to an assessment of the management qualities of the leadership.
- This paper *establishes* a profile for strategy implementation.
- Three different computational schemes *are developed*.
- The author of this paper *proposes* an approach to the creation of an integrated method of investigating and designing objects, based on a local computer system.
- A method *is proposed* for forecasting the night dream analytically with a lead 48 hr.
- *Attention is being paid to* management of environmental effluents and disposal of hazardous wastes in this paper.
- Some of the specific topics *discussed* in the paper are...
- *Included in this paper are* also the measuring results before and after the following tests...

3. Introduction

The purpose of the Introduction should be to supply sufficient background information to allow the reader to understand and evaluate the results of the present study without needing to refer to previous publications on the topic. The Introduction should also provide the rationale for the present study. Above all, you should state briefly and clearly your purpose in writing the paper. Choose references carefully to provide the most important background information. Much of the Introduction should be written in the present tense, because you will be referring primarily to your problem and the established knowledge relating to it at the start of your work.

Suggested rules for a good Introduction are as follows:

(1) The Introduction should present first, with all possible clarity, the nature and scope of the problem investigated.

It is important to keep in mind, however, that the purpose of the Introduction is to introduce (the paper). Thus, the first rule (definition of the problem) is the cardinal one. And, obviously, if the problem is not stated in a reasonable, understandable way, readers will have no interest in your solution. Even if the reader labors through your paper, which is unlikely if you haven't presented the problem in a meaningful way, he or she will be unimpressed with the brilliance of your solution. In a sense, a scientific paper is like other types of journalism. In the Introduction you should have a "hook" to gain the reader's attention. Why did you choose that subject, and why is it important?

(2) It should review the pertinent literature to orient the reader.

(3) It should state the method of the investigation. If deemed necessary, the reasons for the choice of a particular method should be stated.

The second and third rules relate to the first. The literature review and choice of method should be presented in such a way that the reader will understand what the problem was and how you attempted to resolve it.

(4) It should state the principal results of the investigation.

The first three rules then lead naturally to the fourth, the statement of principal results and conclusions, which should be the capstone of the Introduction. This road map from problem to solution is so important that a bit of redundancy with the Abstract is often desirable.

(5) It should state the principal conclusion(s) suggested by the results.

Do not keep the reader in suspense; let the reader follow the development of the evidence. An O. Henry surprise ending might make good literature, but it hardly fits the mold of the scientific method. Many authors, especially beginning authors, make the mistake (and it is a mistake) of holding back their most important findings until late in the paper. In extreme cases, authors have sometimes omitted important findings from the Abstract, presumably in the hope of building suspense while proceeding to a well-concealed, dramatic climax. However, this is a silly gambit that, among knowledgeable scientists, goes over like a double negative at a grammarians' picnic. Basically, the problem with the surprise ending is that the readers become bored and stop reading long before they get to the punch line. "Reading a scientific article isn't the same as reading a detective story. We want to know from the start that the butler did it" (Ratnoff, 1981).

Citations and Abbreviations

If you have previously published a preliminary note or abstract of the work, you should mention this (with the citation) in the Introduction. If closely related papers have been or are about to be published elsewhere, you should say so in the Introduction, customarily at or toward the end. Such references help to keep the literature neat and tidy for those who must search it.

In addition to the above rules, keep in mind that your paper may well be read by people outside your narrow specialty. Therefore, the Introduction is the proper place to define any specialized terms or abbreviations that you intend to use. Let me put this in context by citing a sentence from a letter of complaint I once received. The complaint was in reference to an ad which had appeared in the Journal of Virology during my tenure as Managing Editor. The ad announced an opening for a virologist at the National Institutes of Health (NIH), and concluded with the statement "An equal opportunity employer, M & F." The letter suggested that "the designation 'M & F' may mean that the NIH is muscular and fit, musical and flatulent, hermaphroditic, or wants a mature applicant in his fifties."

Useful Expressions and Patterns forIntroduction

- Much research has been carried out to...
- Recently, some interesting approach has been suggested/proposed...
- Recently, some interesting approaches for transmitting diversity *have been suggested*.
- The issue of anxiety in L2 learning *has been widely recognized* for its significant impact on the L2 learners.
- Supply chain management *has been hailed* in the popular and academic presses as a cost saving and value-creating process.
- The effective management of information systems (IS) projects continues *to be* a critical organizational challenge and imperative, particularly in order to ensure integration between information technologies and business priorities and activities.
- Very little is known about...
- Very little has been written about...
- Very little is understood about...
- Little has been published on...
- There is, however, very little literature that reveals...
- Research is still needed on...
- ... research has largely ignored...
- The aim of this paper is to discuss/examine/investigate...
- The goal of this paper is to answer/discuss/investigate...
- The purpose in this paper is to discuss...
- It is the goal to try to reveal...

- This paper describes...
- This article investigates...
- This paper shows an approach to...
- The genesis of this research project was to investigate...
- The purpose of this study is to examine...
- This study presents an attempt to provide...
- The current study attempts to describe...
- The purpose of our work in progress is to gain...
- Toward this goal, this study looked closely at...
- In this study, it is concerned to show that...
- In this article, attention was given to show that...

4. Body

Contrary to popular belief, you shouldn't start the Results section by describing methods that you inadvertently omitted from the Materials and Methods section.

There are usually two ingredients of the Results section. First, you should give some kind of overall description of the experiments, providing the "big picture," without, however, repeating the experimental details previously provided in Materials and Methods section. Second, you should present the data. Your results should be presented in the past tense.

Of course, it isn't quite that easy. How do you present the data? A simple transfer of data from laboratory notebook to manuscript will hardly do. Most importantly, in the manuscript you should present representative data rather than endlessly repetitive data. The fact that you could perform the same experiment 100 times without significant divergence in results might be of considerable interest to your major professor, but editors, not to mention readers, prefer a little bit of predigestion. Aaronson (1977) said it another way: "The compulsion to include everything, leaving nothing out, does not prove that one has unlimited information; it proves that one lacks discrimination." Exactly the same concept, and it is an important one, was stated almost a century earlier by John Wesley Powell, a geologist who served as President of the American Association for the Advancement of Science in 1888. In Powell's words: "The fool collects facts; the wise man selects them."

How to Handle Numbers

If one or only a few determinations are to be presented, they should be treated descriptively in the text. Repetitive determinations should be given in tables or graphs.

Any determinations, repetitive or otherwise, should be meaningful. Suppose that, in a particular group of experiments, a number of variables were tested (one at a time, of course). Those variables that affect the reaction become determinations or data and, if extensive, are tabulated or graphed. Those variables that do not seem to affect the reaction need not be tabulated or presented; however, it is often important to define even the negative aspects of your experiments. It is often good insurance to state what you did **not** find under the conditions of your experiments. Someone else very likely may find different results under different conditions.

If statistics are used to describe the results, they should be meaningful statistics.

Strive for Clarity

The results should be short and sweet, without verbiage. Mitchell (1968) quoted Einstein as having said, "If you are out to describe the truth, leave elegance to the tailor." Although the Results section of a paper is the most important part, it is often the shortest, particularly if it is preceded by a well-written Materials and Methods section and followed by a well-written Discussion.

The Results need to be clearly and simply stated because it is the Results that constitute the new knowledge that you are contributing to the world. The earlier parts of the paper (Introduction, Materials and Methods) are designed to tell why and how you got the Results; the later part of the paper (Discussion) is designed to tell what they mean. Obviously, therefore, the whole paper must stand or fall on the basis of the Results. Thus, the Results must be presented with crystal clarity.

Avoid Redundancy

Do not be guilty of redundancy in the Results. The most common fault is the repetition in words of what is already apparent to the reader from examination of the figures and tables. Even worse is the actual presentation, in the text, of all or many of the data shown in the tables or figures. This grave sin is committed so frequently that I comment on it at length, with examples, in the chapters on how to prepare the tables and illustrations.

EXAMPLE 17-6 "It is clearly shown in Table 1 that nocillin inhibited the growth of N. gonorrhoeae."

Analysis Do not be verbose in citing figures and tables. It's better to say "Nocillin inhibited the growth of N. gonorrhoeae (Table 1)."

Textual Coherence Through Exemplification Signals

- An *illustration/example* of this might be...
- This is best *illustrated/exemplified* by...
- Body movements have meaning. *For example/For instance*, the lifting of an eyebrow for disbelief.
- Facial expressions can show characteristics *such as* fear and arrogance…
- Body movements have meaning. *If* the speaker shrugs his shoulders, it might indicate indifference.

Textual Coherence Through Comparison and Contrast Signals

- Skinner is *as* interested *as* Watson in changing behavior.

- Like Watson, Skinner believes in changing behavior.
- Skinner is similar to Watson in that they believe in changing behavior.
- Both men believe that behavior can be changed.
- There are several similarities between Skinner and Watson.
- Skinner resembles Watson in that they both believe in changing behavior.
- Skinner, too, believes in changing behavior through conditioning.
- Watson explained behavior in terms of stimulus-response, and so does Skinner.
- Watson did not focus on behavior which could not be measured, and neither does Skinner.
- Just as Watson believed in stimulus-response, so Skinner is interested primarily in changing behavior through conditioning.
- Watson believed in stimulus-response. Similarly/Likewise/Correspondingly/In the same way/In like manner, Skinner is interested primarily in changing behavior through conditioning.
- The British system is less stable than the American system.
- As opposed to/In contrast to/Unlike the American system, the British system has a plural executive.
- The British system contrasts with /differs from the American in that it has a plural executive.
- There are many differences /contrasts between the two systems.
- The British system is different from/dissimilar to the American system.
- The British system has a plural executive, but the American system…
- The British system has a plural executive. The American system, however/in contrast/ on the other hand, has a singular executive.
- While/Whereas the British system has a plural executive; the American system has a singular one.

Textual Coherence Through Division and Classification

- There are three /several types / kinds of managerial skills.
- Managerial skills can be classified/divided into three types.
- The first category is composed of /is comprised of/consists of necessary techniques.
- There are three types of managerial skills.
- Third /Finally /In addition/Then, there are conceptual skills.
- The third category/Still another type/The final kind is conceptual skills.
- technical, human, and conceptual skills
- (1) technical, (2) human, and (3) conceptual skills
- (a) technical, (b) human, and (c) conceptual skills
- technical skills, human skills, and conceptual skills

Textual Coherence Through Chronology Signals

- (action or time), then/at that point / after that/afterwards/thereafter/before that/ meanwhile /during this time (action).
- While/Before/After/When / By the time/As (action).
- In ancient time, on march 1, in 1864, at the end of the war
- A period, a century, a decade
- Pacioli felt the need for keeping records of business transactions, and he decided to describe the technique of double entry bookkeeping.
- Having felt the need for keeping records of business transactions, Pacioli decided to describe the technique of double entry bookkeeping.

Textual Coherence Through Process Signals

- To insure a smooth-running alibi, a person should start feigning illness at least one day prior to calling in sick.
- It is your day, and it's just begun.
- The first part of your mission has been accomplished.
- Finally, before you leave the office make sure that all of your work is caught up.
- You are now free to have an enjoyable evening, along or with a friend.
- A person that has a truly sound illness alibi still has some follow-up work to do.
- Before you turn off into the land of Nod, set your alarm clock for just a few minutes till starting time for work.
- In the final stage, don't let your conscience bother you when co-workers say with great sympathy that maybe you should not have come back to work so soon.

Textual Coherence Through Causality Signals

- A cause of, a reason for, a result /consequence of
- Organizations cause/lead to/create/result in/produce/contribute to communication barriers.
- Communication barriers result from/stem from the formal structures of organizations.
- to be responsible for, a causal relationship, a resultant condition, a contributing factor
- Organizations create barriers, for they have formal structures.
- Organizations have formal structures, so they create barriers.
- …formal structures; therefore/consequently/as a result/because of this/for this reason/hence, they create barriers.
- Because/Since organizations have formal structures, they create barriers.
- There are so many layers that the message gets distorted.
- The fact that they have formal structures explains why they create barriers.
- The existence of many layers means that messages tend to get distorted.
- As messages have to pass through many layers, they tend to get distorted.

- If messages have to pass through many layers, they tend to get distorted.
- Messages which pass through many layers tend to get distorted.
- The more layers there are, the more messages get distorted.
- Having to pass through many layers, messages tent to get distorted.

Textual Coherence Through Spatial Signals

- The edge, the center, the lower stratum
- …the tiny fibers that extend/stretch/move/run outward from…
- The short fibers branching out around /from/through/outside the cell body…
- South/north, top/bottom, right/left, internal/external
- …the axons that run from the brain to the base of the spinal cord can sometimes be as long as 3 feet, but most axons are only an inch or two in length.

Textual Coherence Through Definition Signals

- Retrieval is the process by which we draw upon the information in memory.
- Motivated forgetting refers to the inability to remember things that we do not want to remember.
- A committee may be defined as any group interacting in regard to a common, explicit purpose with formal authority delegated from and appointing executive.
- Computer programmers (people who write instructions for computers) and system analysts (people who analyze problems and recommend solutions) designed and implemented the applications by studying the problems, the same task that clerks previously performed.
- Those psychologists who have applied basic knowledge in the subdisciplines of motivation, learning, and personality to an understanding of work behavior in organization have identified themselves as industrial psychologists.
- The function of memory is retention, or holding on to events and information from the past.
- When we use the phrase "systematic study," we mean looking at relationships, attempting to attribute causes and effects, and basing our conclusions on scientific evidence—that is, data gathered under controlled conditions and measured and interpreted in a reasonably rigorous manner.
- If previous information or experience interferes with the retrieval of something we have learned more recently, we speak of proactive inhibition…Retroactive inhibition, on the other hand, is caused by items that have been learned after what we are trying to remember.
- A recent development in world business is the multinational or global company—a corporation which maintains world headquarters in one country but performs production, marketing, finance, and personnel functions within many nations.
- Overlapping somewhat with the interests of industrial psychologists are the concerns of

industrial sociologists, who have applied basic knowledge in the field of sociology to an understanding of the behavior of individuals in formal and informal groups.

Useful Expressions and Patterns for Graphs

- A data graph/chart/diagram/illustration/table; pie chart; bar chart/histogram; line chart/curve diagram; table; flow chart/sequence diagram; processing/procedure diagram
- Rapid/rapidly
- Dramatic/dramatically
- Significant/significantly
- Sharp/sharply
- Steep/steeply
- Steady/steadily
- Gradual/gradually
- Slow/slowly
- Slight/slightly
- Stable/stably
- Significant changes
- Noticeable trend
- During the same period
- Grow/grew
- Distribute
- Unequally
- In the case of adv.
- In terms of/in respect of/regarding
- In contrast
- Government policy
- Market forces
- Measure
- Forecast
- Show/describe/illustrate/can be seen from
- Clear/apparent
- Reveal/represent
- Figure/statistic/number
- Percentage/proportion
- The table shows the changes in the number of…over the period from… to…
- The bar chart illustrates that…
- The graph provides some interesting data regarding…
- The diagram shows (that)…
- The pie graph depicts (that)…

- This is a curve graph which describes the trend of…
- The tree diagram reveals how…
- The data/statistics show (that)…
- The data/statistics/figures lead us to the conclusion that…
- As is shown/demonstrated/exhibited in the diagram/graph/chart/table…
- According to the chart/figures…
- As is shown in the table…
- As can be seen from the diagram, great changes have taken place in…
- From the table/chart/diagram/figure, it can be seen clearly that…or it is clear/apparent from the chart that…
- This is a graph which illustrates…
- This table shows the changing proportion of a & b from…to…
- The graph, presented in a pie chart, shows the general trend in…
- This is a column chart showing…
- As can be seen from the graph, the two curves show the fluctuation of…
- Over the period from…to…the…remained level.
- In the year between…and…
- In the 3years spanning from 1995 through 1998…
- From then on/from this time onwards…
- The number of…remained steady/stable from (month/year) to (month/year).
- The number sharply went up to…
- The percentage of …stayed the same between… and…
- The figures peaked at… in (month/year)
- The percentage remained steady at…
- The percentage of…is slightly larger/smaller than that of…
- There is not a great deal of difference between…and…
- The graphs show a threefold increase in the number of…
- …decreased year by year while…increased steadily.
- The situation reached a peak (a high point at…) of [%]
- The figures/situation bottomed out in…
- The figures reached the bottom/a low point/hit a trough.
- A is…times as much/many as B.
- A increased by…
- A increased to…
- There is an upward trend in the number of…
- A considerable increase/decrease occurred from…to…
- From…to…the rate of decrease slow down.
- From this year on, there was a gradual declined reduction in the…, reaching a figure of…
- Be similar to…
- Be the same as…

- There are a lot similarities/differences between…and…
- A has something in common with B
- The difference between A and B lies in…
- …(year) witnessed/saw a sharp rise in…

5. Conclusion

Useful Expressions and Patterns for Conclusion

- The data; this study; the results; this comparison; the observation; the new findings; this work, etc.
- Provide; indicate; reveal; present; throw light on; expose; illustrate; examine; suggest; identify, etc.
- The results described here are compatible with what one could have expected on the basis of the general understanding of vesicle information.
- The findings are roughly consistent with…
- The data presented here offer little or no support for the idea that…
- These findings seem to contradict the view that. .
- The results is consistent with the view that…
- The findings are coherent with…
- This result corresponds to the view that…
- The data provided by this project has revealed that…
- This work presents evidence of a more complex role for…
- The results presented here provide a new structural basis to further experiments required for a detailed understanding of the complex mechanism of…
- This work provides new sight into…
- The new findings provide a clearer picture of this rapidly and chaotically evolving system.
- This article has gone beyond definition and description of religion to examine the question of the uniqueness of religion on empirical grounds.
- This article has offered a different possibility, that there is something unique about religion in and of itself.
- It should be noted that this study has examined only…
- This analysis has concentrated on…
- The findings of this study are restricted to…
- This study has addressed only the question of…
- The limitations of this study are clear…
- It should be pointed out that…has/have not…
- The model assumes that…does not affect…
- The study supposes that…is at a fixed level for a given period. However, in practice…
- …is ignored
- However, the findings do not imply…

- The results of this study cannot be taken as evidence for…
- Unfortunately, it cannot be determined from the data…
- The lack of…means that it can not be certain that…
- These cases illustrate the need to research…
- …proves to be a fruitful direction for future research.
- …warrants further investigation
- Further studies, involving… are expected to yield…
- …remains to be determined.
- Replication of this study using…may further elucidate…
- It is expected that further analysis of this data by the research community will be valuable in understanding…

6. Acknowledgments

Two possible ingredients require consideration. First, you should acknowledge any significant technical help that you received from any individual, whether in your laboratory or elsewhere. You should also acknowledge the source of special equipment, cultures, or other materials. You might, for example, say something like "Thanks are due to J. Jones for assistance with the experiments and to R. Smith for valuable discussion." (Of course, most of us who have been around for a while recognize that this is simply a thinly veiled way of admitting that Jones did the work and Smith explained what it meant.) Second, it is usually the Acknowledgments wherein you should acknowledge any outside financial assistance, such as grants, contracts, or fellowships. (In these days, you might snidely mention the absence of such grants, contracts, or fellowships.)

Being Courteous

The important element in Acknowledgments is simple courtesy. There isn't anything really scientific about this section of a scientific paper. The same rules that would apply in any other area of civilized life should apply here. If you borrowed a neighbor's lawn mower, you would (I hope) remember to say thanks for it. If your neighbor gave you a really good idea for landscaping your property and you then put that idea into effect, you would (I hope) remember to say thank you. It is the same in science; if your neighbor (your colleague) provided important ideas, important supplies, or important equipment, you should thank him or her. And you must say thanks in print, because that is the way that scientific landscaping is presented to its public.

A word of caution is in order. Often, it is wise to show the proposed wording of the Acknowledgment to the person whose help you are acknowledging. He or she might well believe that your acknowledgment is insufficient or (worse) that it is too effusive. If you have been working so closely with an individual that you have borrowed either equipment or ideas, that person is most likely a friend or a valued colleague. It would be silly to risk either your friendship or the opportunities for future collaboration by placing in public print a thoughtless word that might be offensive. An inappropriate thank you can be worse than none at all, and if you value the advice and help of friends and colleagues, you should be careful to thank them in a way that pleases rather than displeases them.

Furthermore, if your acknowledgment relates to an idea, suggestion, or interpretation, be very specific about it. If your colleague's input is too broadly stated, he or she could well be placed in the sensitive and embarrassing position of having to defend the entire paper. Certainly, if your colleague is not a coauthor, you must not make him or her a responsible party to the basic considerations treated in your paper. Indeed, your colleague may not agree with some of your central points, and it is not good science and not good ethics for you to phrase the Acknowledgments in a way that seemingly denotes endorsement.

EXAMPLE 17-7 "I wish to thank John Jones."

Analysis Do not be verbose in citing figures and tables. Say "Nocillin inhibited the growth of N. gonorrhoeae (Table 1)." The word "wish" would disappear from Acknowledgments. Wish is a perfectly good word when you mean wish, as in "I wish you success." However, if you say "I wish to thank John Jones," you are wasting words. You may also be introducing the implication that "I wish that I could thank John Jones for his help but it wasn't all that great." "I thank John Jones" is sufficient.

Be thoughtful, analytical, and critical about your data and ideas. Figure out what is novel in what you did. Remember that that there are few data sets so imbued with novelty that they can't be made dull, and few that are so dull that there aren't novel insights that can be drawn from them. It is your job to find the novelty and highlight it. If you've found the novelty, you've done the hart part—nature gives up her secrets grudgingly. We all wrestle with our data sets, trying to figure out their meaning and their story.

It's only after this that specific language skills mater. You must produce a document in which, at an absolute minimum, the right words are used, they are spelled correctly, and the rules of grammar and usage are followed. It is your responsibility as the author to ensure this. Do not submit a manuscript thinking that the reviewers, the editors, or the publisher will fix imperfect English. They won't. It isn't their job, none of them have the time, and the journals don't have the money. Most journals screen papers for language and bounce back those that are not up to an acceptable standard; they won't send them out for review. They have a responsibility not to overwork reviewers by sending them papers that are not ready. The author's job is to make the reader's job easy. Actually, many of the journals do help with language and writing. They know that beginning writers struggle, and most of them want to help. But they usually do so when it means tidying up and fixing quirks of English, rather than doing a full copy edit. It is also an act of generosity you should not count on. Editors help those who help themselves.

The tool most authors rely on to fix writing problems is their word processor. The spell checker is essential, but it will miss errors like "their" versus "there" and typos that create a real but wrong word, like "from" versus "form". Then there is the grammar checker; this can be useful in catching some errors and it's better than nothing (but not much). As you write, you can periodically check on the things it underlines—it catches some real errors, but it makes a lot of mistakes.

Better information is available in any of a number of excellent books and websites. For those who are native English speakers and experience writers, they still have a shell full of books on grammar and language and keep a bookmark in the Web browser to the Oxford English Dictionary. It is essential to have good references. Countless books have been written for people who are insecure in their knowledge of English. For guides to grammar and usage, shorter is better. You don't need to understand the deepest arcana of English grammar—you need practical, everyday advice. It's no accident that the most battered and coffee-stained book on every writer's shelf is the shortest: Strunk and White, The Elements of Style (the original 1918 version by Strunk is available online for free, http://www.bartleby.com/141/).

The advice most people will give you, however, is not a reference book, but to give your manuscript to an English-speaking colleague to go over before you submit. This can be useful, but it is recommended not to rely on a friend down the hall as your only language check—at least, not unless they are both a good friend and a good editor. The editors have sent back too many papers that were edited by friends who hadn't done an adequate job, and have had some "polite disagreements" with authors who were sure that because their American friend looker over the paper it must be okay. Editing is difficult and time-consuming. Most friends don't have the time, and many don't have the skills, to do a complete and careful word-by-word edit. There are professional services that do this; some are excellent, and they aren't very expensive. Some publishers list editing services on their websites. After spending the equivalent thousands of dollars to do the research, spending a few hundred more to ensure the final paper is of the highest possible caliber is a small and worthwhile investment. When you need the job done well, use a skilled professional.

My suggestion to not rely on an English-speaking colleague changes completely, however, when that colleague is a coauthor. All authors are responsible for a paper's entire content, and that includes the language. Your English-speaking coauthor is responsible for ensuring the language is correct. When reviewers read poorly written papers with coauthors from the United States, Great Britain, or other English-speaking countries, they can be appropriately brutal. They may question whether those authors were actually involved in the paper or whether they merely failed in their responsibility to ensure it was ready to submit. Either way, your coauthor doesn't look good. Unfortunately, as fallout, you may not look good either. If you are collaborating with a native English speaker, make sure he or she will be willing to do the necessary language-editing, and make sure you allow appropriate time to do it.

As a closing story, a colleague of mine questioned whether this book would be useful for scholars for whom English is a second language. She worried that for people who struggle to write grammatical sentences, my focus on storytelling might be overskill. I pointed out that as an editor, when I get a paper where the story is strong but the language weak, I'll send it back to get the language fixed before sending it out for review. If I get a paper where the story is weak I'll just reject it.

So which is more important—getting the grammar or the story down? I'll vote for story every time. You can hire an editor to help with the language. But you can't hire a scientist to help with the science. It's your science and only you can develop the story. Remember, always, that science is not about information; it is about knowledge and understanding. If you can offer understanding, you are most of the way to writing a paper that will be publishable in the world's best journals.

Useful Expressions and Patterns for Acknowledgement

- The authors gratefully acknowledge this support.
- …acknowledge the contributions made by…
- …acknowledge the assistance of …in…
- …acknowledge the contributions of …in…
- …acknowledge the comment and issues raised by…on…for…
- …acknowledge…who…
- The contributions of/made by… are gratefully acknowledged.
- …thank…for…
- …be grateful/indebted to…for…
- …special thanks go to…for…
- …record…thanks to…who…
- …thank…who…
- …owe much to…who…
- Much indebtedness is owed to…

17.2 How to Submit a Scientific Paper

1. Choosing a Journal

Science isn't complete until it has been published, and the first step in that process is identifying your audience and choosing a journal to submit the paper to.

The pressure to be relevant can lead to studies that provide information useful to local managers or industries but may not offer knowledge that would be relevant to a global audience. The pressure to succeed, however, can lead researchers to submit those papers to high-profile journals even when they may not be a good fit. Many papers were rigorously done but only offer information. In these papers, authors often highlight that what is novel is that it presents the first data on a process in a new region—trace gas emissions, nitrification, and so on. But the paper was rejected.

Any leading journal is likely to reject a paper if all it does is flesh out the information base: it's the first data set on a new region, it demonstrates that a reaction works similarly with a slightly different substitution pattern on a molecule, or that the gene sequence from a new bacterium is only modestly different from that in known bacteria.

This isn't about basic versus applied research. It's about information versus knowledge. First-rate applied research goes beyond presenting a data set—it provides broader insights into the nature of the problem, insights that are useful to people working on related problems and in different areas. For example, a paper on how plowing a soil alters nitrate leaching and nitrous oxide emissions might be valuable for local managers who are trying to maximize crop yield while minimizing groundwater pollution and greenhouse gas emissions. They need the information, and it should be published in an appropriate venue. But unless the paper also offers new insights into the fundamentals of N-cycling or develops a new, transferable management regime, that venue is not likely to be a high-impact basic-science journal—and that will be true regardless of whether the work was done in India or Indiana.

So before you submit, make sure you know a journal's focus and intended audience. Do you want to offer local farmers improved tillage techniques or soil biologists new insights into how bacteria process N? Read a journal's description carefully and analyze the papers it publishes. If you are still unsure, email the editor and ask for advice. Then pick a journal appropriate for your story and intended audience. Don't focus on the journal's status, but on its scope. There will always be a draw toward the journals with the highest impact, but submitting a paper that doesn't fit is a waste of everyone's time and energy. Ultimately, journal prestige means little—the top journals publish some mediocre papers and lower impact journals publish some extraordinary ones. In the modern world of search engines and open-access journals, good papers will be found and cited whereas bad ones will be ignored, regardless of where they are published.

2. Checking Your Manuscript

- Make sure you have totally followed the Instructions to Author from the chosen journal which usually supplies instructions on their websites.
- Unless the journal (or the style manual it says to use) instructs otherwise follow these guidelines:

  Double space.

  Use margins of at least 1 inch (at least about 25 mm).

  Start each section of the manuscript on a new page. Figure legends are grouped on a separate page. Normally, the tables, figures, and the figure legends should be assembled at the back of the manuscript. (Though, some journals asked authors to insert them in the text.)
- Give your manuscript to an English-speaking friend or colleague to go over before you submit. This can be useful, but it is recommended not to rely on a friend down the hall as your only language check—at least, not unless they are both a good friend and a good editor. The editors have sent back too many papers that were edited by friends who hadn't done an adequate job, and have had some "polite disagreements" with authors who were sure that because their American friend looker over the paper it must be

okay. Editing is difficult and time-consuming. Most friends don't have the time, and many don't have the skills, to do a complete and careful word-by-word edit.

- There are professional services that do this; some are excellent, and they aren't very expensive. Some publishers list editing services on their websites. After spending the equivalent thousands of dollars to do the research, spending a few hundred more to ensure the final paper is of the highest possible caliber is a small and worthwhile investment. When you need the job done well, use a skilled professional.
- In addition to proofreading the manuscript yourself, try to have someone do so who has not seen the manuscript before and so may notice problems that you miss.

3. Cover Letter

It is worth noting that you should always send a cover letter with the manuscript. Manuscripts without cover letters pose immediate problems: To which journal is the manuscript being submitted? Is it a new manuscript, a revision requested by an editor (and, if so, which editor?), or a manuscript perhaps misdirected by a reviewer or an editor? If there are several authors, which one should be considered the submitting author, at which address? The address is of special importance, because the address shown on the manuscript may not be the current address of the contributing author. The contributing author should also include his or her telephone number, e-mail address, and fax number in the cover letter or on the title page of the manuscript. It is often helpful to suggest the appropriate editor (in multieditor journals) and possible reviewers. Some journals also let authors list people they believe should not review their manuscripts, for example because of conflicts of interest.

If not obvious, state the section of the journal that the paper is intended for or the category of article being submitted. Also provide any other information that the Instructions for Authors say to include. Be kind to the editor and you might even choose to say something nice.

More and more journals are requiring that manuscripts be submitted electronically, through Web sites that they designate. If you are submitting your manuscript electronically, the manuscript submission Web site may supply a mechanism for providing your cover letter. Alternatively, it may prompt you for the information the journal wants to receive, thus automatically generating a cover letter or the equivalent. This electronic option saves you the trouble of composing a letter and helps ensure that the journal receives the information it requires.

Sample Cover Letter

Dear Dr. ________:

Enclosed are two complete copies of a manuscript by Mary Q. Smith and John L. Jones titled "Fatty Acid Metabolism in Cedecia neteri," which is being submitted for possible publication in the Physiology and Metabolism section of the Journal of Bacteriology.

This manuscript is new, is not being considered elsewhere, and reports new findings that extend results we reported earlier in The Journal of Biological Chemistry (145: 112-117, 1992). An abstract of this manuscript was presented earlier (Abstr. Annu. Meet. Am. Soc. Microbiol., p. 406, 1993).

Sincerely,

Mary Q. Smith

Useful Expressions and Patterns

- We would be glad if our manuscript would give you complete satisfaction.
- We deeply appreciate your consideration of our manuscript, and we look forward to receiving comments from the reviewers.

4. Follow-up Correspondence

Most journals send out an "acknowledgment of receipt" by e-mail or other means when the manuscript is received. If you do not receive an acknowledgment in 2 weeks (or less for electronically submitted manuscripts), call or write the editorial office to verify that your manuscript was indeed received. There was one author whose manuscript was lost in the mail, and it was not until 9 months later that the problem was brought to light by his meek inquiry as to whether the reviewers had reached a decision about the manuscript.

Busy editors and reviewers being what they are, do not be concerned if you do not receive a decision within one month after submission of the manuscript. Most journal editors, at least the good ones, try to reach a decision within 4 to 6 weeks or, if there is to be further delay for some reason, provide some explanation to the author. If you have had no word about the disposition of your manuscript after 8 weeks have elapsed, it is not at all inappropriate to send a courteous inquiry to the editor.

Useful Expressions and Patterns

- I would be greatly appreciated if you could spend some of your time check the status for us. I am very pleased to hear from you on the reviewer's comments.
- I am not sure if it is the right time to contact you to inquire about the status of my submitted manuscript titled "Paper Title".
- We tried our best to improve the manuscript and made some changes in the manuscript.
- I am just wondering that my manuscript has been sent to reviewers or not?
- We are very sorry for our negligence of…
- We are very sorry for our incorrect writing…
- It is really true as reviewer suggested that…
- We have made correction according to the reviewer's comments.
- We have re-written this part according to the reviewer's suggestion.
- As reviewer suggested that…
- Considering the Reviewer's suggestion, we have…

5. The Editor's Decision

The editor's decision will be one of three general types, commonly expressed in one word as "accept," "reject," or "modify." Normally, one of these three decisions will be reached within 4 to 6 weeks after submission of the manuscript. If you are not advised of the editor's decision within 8 weeks, or provided with any explanation for the delay, do not be afraid to write to or call the editor. You have the right to expect a decision, or at least a report, within a reasonable period of time; also, your inquiry may bring to light a problem. Obviously, the editor's decision could have been made but notification did not reach you. If the delay was caused within the editor's office (usually by lack of response from one of the reviewers), your inquiry is likely to trigger an effort to resolve the problem, whatever it is.

Besides which, you should never be afraid to talk to editors. With rare exceptions, editors are awfully nice people. Never consider them adversaries. They are on *your* side. Their only goal is to publish good science in understandable language. If that is not your goal also, you will indeed be dealing with a deadly adversary; however, if you share the same goal, you will find the editor to be a resolute ally. You are likely to receive advice and guidance that you could not possibly buy.

More likely, you will receive from the editor a covering letter and two or more lists labeled "reviewers' comments." The letter may say something like "Your manuscript has been reviewed, and it is being returned to you with the attached comments and suggestions. We believe these comments will help you improve your manuscript." This is the beginning phraseology of a typical modify letter. The letter may go on to say that the paper will be published if modified as requested, or it may say only that it will be reconsidered if the modifications are made.

By no means should you feel disconsolate when you receive such a letter. Realistically, you should not expect that rarest of all species, the accept letter without a request for modification. The vast majority of submitting authors will receive either a modify letter or a reject letter, so you should be pleased to receive the former rather than the latter. When you receive a modify letter, examine it and the accompanying reviewers' comments carefully. (In many cases, the modify letter is a form letter, and it is the accompanying comments that are significant, for example regarding a point about which the reviewers disagree.) The big question now is whether you can, and are willing to, make the changes requested.

If both referees point to the same problem in a manuscript, almost certainly it *is* a problem. Occasionally, a referee may be biased, but hardly two of them simultaneously. If referees misunderstand, readers will. Thus, my advice is: If two referees misunderstand the manuscript, find out what is wrong and correct it before resubmitting the manuscript to the same journal or another journal.

If the requested changes are relatively few and slight, you should go ahead and make them. As King Arthur used to say, "Don't get on your high horse unless you have a deep moat to cross." If major revision is requested, however, you should step back and take a total look at your position. One of several circumstances is likely to exist.

First, the reviewers are right, and you now see that there are fundamental flaws in your paper. In that event, you should follow their directions and rewrite the manuscript accordingly.

Second, the reviewers have caught you off base on a point or two, but some of the criticism is invalid. In that event, you should rewrite the manuscript with two objectives in mind: Incorporate all of the suggested changes that you can reasonably accept, and try to beef up or clarify those points to which the reviewers (wrongly, in your opinion) took exception. Finally, and importantly, when you resubmit the revised manuscript, provide a covering statement indicating your point by point what you did about the reviewers' comments.

Third, it is entirely possible that one or both reviewers and the editor seriously misread or misunderstood your manuscript, and you believe that their criticisms are almost totally erroneous. In that event, you have two alternatives. The first, and more feasible, is to submit the manuscript to another journal, hoping that your manuscript will be judged more fairly. If, however, you have strong reasons for wanting to publish that particular manuscript in that particular journal, do not back off; resubmit the manuscript. In this case, however, you should use all of the tact at your command. Not only must you give a point-by-point rebuttal of the reviewers' comments; you must do it in a way that is not antagonistic. Remember that the editor is trying hard, probably without pay, to reach a scientific decision. If you start your covering letter by saying that the reviewers, whom the editor obviously has selected, are "stupid" (I have seen such letters), I will give you 100 to 1 that your manuscript will be immediately returned without further consideration. On the other hand, every editor knows that *every* reviewer can be wrong and in time (Murphy's law) will be wrong. Therefore, if you calmly point out to the editor exactly why you are right and the reviewer is wrong (never say that the editor is wrong), the editor is very likely to accept your manuscript at that point or, at least, send it out to one or more additional reviewers for further consideration.

If you do decide to revise and resubmit the manuscript, try very hard to meet whatever deadline the editor establishes. Most editors do set deadlines. Obviously, many manuscripts returned for revision are not resubmitted to the same journal; hence, the journal's records can be cleared of deadwood by considering manuscripts to be withdrawn after the deadline date passes.

If you meet the editor's deadline, he or she may accept the manuscript forthwith. Or, if the modification has been substantial, the editor may return it to the same reviewers. If you have met, or defended your paper against, the previous criticism, your manuscript will probably be accepted. On the other hand, if you fail to meet the deadline, your revised manuscript may be treated as a new manuscript and again subjected to full review, possibly by a different set of reviewers. It is wise to avoid this double jeopardy, plus additional review time, by carefully observing the editor's deadline if it is at all possible to do so. If you believe that you cannot meet the deadline, immediately explain the situation to the editor; the deadline might then be extended.

Useful Expressions and Patterns

- Thank you for your detailed and lengthy criticism of my manuscript. I will be sure to incorporate your suggestions in my next draft.
- On behalf of my co-authors, we thank you very much for giving us an opportunity to revise our manuscript, we appreciate editor and reviewers very much for their positive and constructive comments and suggestions on our manuscript entitled "Paper Title".
- We would like to express our great appreciation to you and reviewers for comments on our paper. Looking forward to hearing from you.
- The main corrections in the paper and the responds to the reviewer's comments are as following…
- Sorry for disturbing you. I am not sure if it is the right time to contact you to inquire about the status of our accepted manuscript titled "Paper Title".
- We greatly appreciate the efficient, professional and rapid processing of our paper by your team. If there is anything else we should do, please do not hesitate to let us know.

Status of submission

Submitted to Journal
Manuscript received by editorial office/Notification of manuscript receipt
With editor
Awaiting editor assignment /editor assigned
Editor declined invitation
Decision letter being prepared
Reviewer(s) invited /with referees
Under review
Required reviews completed
Decision in process
Minor revision/Major revision
Accepted
Rejected
Transfer copyright form
Uncorrected proof/Galley proof
In Press, corrected proof
Manuscript sent to Pproduction
In production

17.3 Attend an International Conference

1. How to Prepare a Poster

1.1 Sizes and Shapes

In recent years, poster displays have become ever more common at both national and international meetings. (Posters are display boards on which scientists show their data and describe their experiments.) As attendance at meetings increased, and as pressure mounted on program committees to schedule more and more papers for oral presentation, something had to change. The large annual meetings, such as those of the Federation of American Societies of Experimental Biology, got to the point where available meeting rooms were simply exhausted. And, even when sufficient numbers of rooms were available, the resulting large numbers of concurrent sessions made it difficult or impossible for attending scientists to keep up with the work being presented by colleagues.

At first, program committees simply rejected whatever number of abstracts was deemed to be beyond the capabilities of meeting room space. Then, as poster sessions were developed, program committees were able to take the sting out of rejection by advising the "rejectees" that they could consider presenting their work as posters. In the early days, the posters were actually relegated to the hallways of the meeting hotels or conference centers; still, many authors, especially graduate students attempting to present their first paper, were happy to have their work accepted for a poster session rather than being knocked off the program entirely. Also, the younger generation of scientists had come of age during the era of science fairs, and they liked posters.

Nowadays, of course, poster sessions have become an accepted and meaningful part of many meetings. Large societies set aside substantial space for the poster presentations. At a recent Annual Meeting of the American Society for Microbiology, about 2,500 posters were presented. Even small societies often encourage poster presentations, because many people have now come to believe that some types of material can be presented more effectively in poster graphics than in the confines of the traditional 10-minute oral presentation.

As poster sessions became normal parts of many society meetings, the rules governing the preparation of posters have become much more strict. When a large number of posters have to be fitted into a given space, obviously the requirements have to be carefully stated. Also, as posters have become common, convention bureaus have made it their business to supply stands and other materials; scientists could thus avoid shipping or carrying bulky materials to the convention city.

Don't ever commence the actual preparation of a poster until you know the requirements specified by the meeting organizers. You of course must know the height and width of the stand. You also must know the approved methods of fixing exhibit materials to the stand. The minimum sizes of type may be specified, and the sequence of presentation may be specified (usually from left to right). This information is usually provided in the program for the meeting.

1.2 Organization

The organization of a poster normally should follow the IMRAD format, although graphic considerations and the need for simplicity should be kept in mind. There is very little text in a well-designed poster, most of the space being used for illustrations.

The Introduction should present the problem succinctly; the poster will fail unless it has a clear statement of purpose right at the beginning. The Methods section will be very brief; perhaps just a sentence or two will suffice to describe the type of approach used. The Results, which is often the shortest part of a written paper, is usually the major part of a well-designed poster. Most of the available space will be used to illustrate Results. The Discussion should be brief. Some of the best posters I have seen did not even use the heading "Discussion"; instead, the heading "Conclusions" appeared over the far-right panel, the individual conclusions perhaps being in the form of numbered short sentences. Literature citations should be kept to a minimum.

1.3 Preparing the Poster

You should number your poster to agree with the program of the meeting. The title should be short and attentiongrabbing (if possible); if it is too long, it might not fit on the display stand. The title should be readable out to a distance of 10 feet (3 m). The typeface should be bold and black, and the type should be about 30 mm high. The names of the authors should be somewhat smaller (perhaps 20 mm). The text type should be about 4 mm high. (A type size of 24 points is suitable for text.) Transfer letters (e. g., Letraset) are an excellent alternative, especially for headings. A neat trick is to use transfer letters for your title by mounting them on standard (2¼-inch) adding machine tape. You can then roll up your title, put it in your briefcase, and then tack it on the poster board at the meeting. Computers can produce display-size type as well.

A poster should be self-explanatory, allowing different viewers to proceed at their own pace. If the author has to spend most of his or her time merely explaining the poster rather than responding to scientific questions, the poster is largely a failure.

Lots of white space throughout the poster is important. Distracting clutter will drive people off. Try to make it very clear what is meant to be looked at first, second, etc. (although many people will still read the poster backwards). Visual impact is particularly critical in a poster session. If you lack graphic talent, consider getting the help of a graphic artist. Such a professional can produce an attractive poster either in the traditional board-mounted style or in the newer single-unit photographic reproduction (superstat).

Robin Morgan, Professor of Animal and Food Sciences at the University of Delaware, told me this: "I'm one of those'science fair"scientists who love posters, and so we make a lot of them. I write text in Word and prepare individual graphics as EPS by using McDraw Pro, DeltaGraph, and Quark. Then, I send the individual parts to a graphic artist. The artist adds a bit of color here and there and lays it all out so it looks good. I then have it printed at a service bureau and have it laminated. The cost is $1,000 per poster (pretty high for many scientists), but it's great to bring home a poster after the meeting and display it in your office or lab."

A poster should contain highlights, so that passersby can easily discern whether the poster is something of interest to them. If they are interested, there will be plenty of time to ask questions about the details. Also, it is a good idea to prepare handouts containing more detailed information; they will be appreciated by colleagues with similar specialties.

A poster may actually be better than an oral presentation for showing the results of a complex experiment. In a poster, you can organize the highlights of the several threads well enough to give informed viewers the chance to recognize what is going on and then get the details if they so desire. The oral presentation, as stated in the preceding chapter, is better for getting across a single result or point.

The really nice thing about posters is the variety of illustrations that can be used. There is no bar (as there often is in journal publication) to the use of color. All kinds of photographs, graphs, drawings, paintings, X-rays, and even cartoons can be presented.

I have seen many excellent posters. Some scientists do indeed have considerable creative ability. It is obvious that these people are proud of the science they are doing and that they are pleased to put it all into a pretty picture. I have also seen many terrible posters. A few were simply badly designed. The great majority of bad posters are bad because the author is trying to present too much. Huge blocks of typed material, especially if the type is small, will not be read. Crowds will gather around the simple, well-illustrated posters; the cluttered, wordy posters will be ignored.

2. How to Present a Paper Orally

2.1 Organization of the Paper

- Starting with "what was the problem?" and ending with "what is the solution?"
  The best way to organize a paper for oral presentation is to proceed in the same logical pathway that one usually does in writing a paper.
- The oral presentation need not and should not contain all of the experimental detail, unless by chance you have been called upon to administer a soporific at a meeting of insomniacs.
  It is important to remember that oral presentation of a paper does not constitute publication, and therefore different rules apply. The greatest distinction is that the published paper must contain the full experimental protocol, so that the experiments can be repeated.
- Extensive citation of the literature is also undesirable in an oral presentation.

2.2 Presentation of the Paper

- You should stick to your most important point or result and stress that.
  Most oral presentations are short (with a limit of 10 minutes at many meetings). Thus, even the theoretical content must be trimmed down relative to that of a written paper. No matter how well organized, too many ideas too quickly presented will be confusing. There will not be time for you to present all your other neat ideas. There

are, of course, other and longer types of oral presentations. A typical time allotted for symposium presentations is 20 minutes. A few are longer. A seminar is normally one hour. Obviously, you can present more material if you have more time.

- You should go slowly and clearly, carefully presenting a few main points or themes.
  If you proceed too fast, especially at the beginning, your audience will lose the thread; the daydreams will begin and your message will be lost. Speak very clearly, and avoid speaking quickly, especially if the language in which you are presenting is not the native language of all the audience members.
- Carefully plan your presentation to fit the allotted time.
  Time limits for conference presentations tend to be strictly enforced. Avoid that you be whisked from the podium before you can report your major result. if possible, make your presentation a bit short (say, 9 or 9.5 minutes if 10 minutes are alloted), to accommodate unexpected slowdowns. Rehearse beforehand, both to make sure it is the right length and to help ensure smooth delivery; stay aware of the time during your presentation. Perhaps indicate in your notes what point in the presentation you should have reached by what time, so that if necessary you can adjust your pace accordingly.
- Remember to look at the audience; avoid habits that might be distracting.
  Sow interest in your subject. Avoid jangling the change in your pocket or repeatedly saying "um" or "you know" or the equivalent from your native language. To polish your delivery, consider videotaping rehearsals of one or more of your presentations.
- Consider following suggestions to handle stage fright.
  (1) Prepare well so you can feel confident, but do not prepare so much that you feel obsessed. (2) To dissipate nervous energy, take a walk or take advantage of the exercise facilities in the conference hotel. (3) Beware of too much caffeine, food, or water. (4) Hide physical signs of anxiety; for example, if your hands tremble under stress, do not hold a laser pointer. (5) Realize that a presentation need not be flawless to be excellent. (6) Perhaps most important, realize that the audience members are there not because they wish to judge your speaking style but because they are interested in your research.
- Avoid (1) talking too rapidly; (2) speaking in a monotone; (3) using too high a vocal pitch; (4) talking and not saying much; (5) presenting without enough emotion or passion; (6) talking down to the audience; (7) using too many "big" words; (8) using abstractions without giving concrete examples; (9) using unfamiliar technical jargon; (10) using slang or profanity; (11) disorganized and rambling performance; (12) indirect communication i. e. beating around the bush.

2.3 Slides

At small, informal scientific meetings, various types of visual aids may be used. Overhead projectors, flip charts, and even blackboards can be used effectively. At most scientific meetings, however, PowerPoint presentations or other slide presentations are the lingua franca. Every scientist should know how to prepare effective slides and use them effectively, yet attendance at almost any meeting quickly indicates that many do not.

- Slides should be designed specifically for use with oral presentations, with large enough lettering to be seen from the back of the room.

  Slides prepared from graphs that were drawn for journal publication are seldom effective and often are not even legible. Slides prepared from a printed journal are almost never effective.
- Slides should not be crowded.

  Each slide should be designed to illustrate a particular point or perhaps to summarize a few. To permit reading, use bullet points, not paragraphs. If a slide cannot be understood in 4 seconds, it is a bad slide.
- A Graph is better than a table.

  If there are findings that you can present in either a graph or a table, use a graph. Doing so will help the audience grasp the point morequickly.
- Indicate the main message in an illustration or table.

  If you show a slide of an illustration or table, indicate its main message. As one long-suffering audience member said, "Don't just point at it."
- Beware of showing too many slides.

  A moderate number of well-chosen slides will enhance your presentation; too many will be distraction.
- Never read the slide text to the audience.

  Normally, each slide should make one simple, easily understood visual statement. The slide should supplement what you are saying at the time the slide is on the screen; the slide should not simply repeat what you are saying. Just reading would be an insult to your audience, unless you are addressing a group of illiterates.

2.4 The Audience

The presentation of a paper at a scientific meeting is a two-way process. Because the material being communicated at a scientific conference is likely to be the newest available information in that field, both the speakers and the audience should accept certain obligations. As indicated above, speakers should present their material clearly and effectively so that the audience can understand and learn from the information being communicated.

Almost certainly, the audience for an oral presentation will be more diverse than the readership of a scientific paper. Therefore, the oral presentation should be pitched at a more general level than would be a written paper. Avoid technical detail. Define terms. Explain difficult concepts. Repeat important points.

Rehearsing a paper before the members (even just a few members) of one's own department or group can make the difference between success and disaster.

The best part of an oral presentation is often the question-and-answer period. During this time, members of the audience have the option, if not the obligation, of raising questions not covered by the speakers, and of briefly presenting ideas or data that confirm or contrast with those presented by the speaker. Such questions and comments should be stated courteously and professionally. This is not the time for some windbag to vent spleen or to describe his or her own erudition in infinite detail. It is all right to disagree, but do not be disagreeable. In short, the speaker has an obligation to be considerate to the audience, and the audience has an obligation to be considerate to the speaker.

What should you do if an audience member is indeed abrasive? If someone asks an irrelevant question? If a question is relevant but you lack the answer?

- If someone is rude, stay calm and courteous. Thank him or her for the question or comment, and if you have a substantive reply, provide it. If the person keeps pursuing the point, offer to talk after the session. You can say, "Well, we'll have to agree to disagree on this point," or "Unfortunately, there is no time to go into this more deeply right now."
- If a question is irrelevant, take a cue from politicians and try to deflect the discussion to something related that you wish to address—perhaps a point you had hoped to include in your presentation but lacked time for. ("That's an interesting question, but a more immediate concern to us was …") Alternatively, offer to talk later. You can also say, "Well, I don't think we have time to discuss that point right now, but I'll be happy to talk with you about it later." "That's an interesting question, but my presentation doesn't really deal with that issue," or "I'm afraid I don't see how that question applies to what I've said."
- If you lack the answer to a question, do not panic—and definitely do not bluff. Admit that you do not know. If you can provide the answer later, offer to do so; if you know how to find the answer, say how. To help prepare for questions that might arise, have colleagues quiz you after your rehearse.
- Instead of asking a question, the person states his or her own viewpoints that agree with yours. You can say, "Yes, that fits in exactly with what I was saying."
- The person states positively that some information you have given is inaccurate, but you are absolutely sure that you are correct. You can provide information source or additional support for your statement. Or you can say, "I believe that my information is correct, but I will certainly recheck my facts." If you are not sure that your information war correct or not, do not feel threatened or regard this as an attack. You can say, "I appreciate your bringing this to my attention. I'll have to recheck my sources to see what is correct.
- The person asked many questions. You can say, "I'm very pleased by your interest, but perhaps we should give others a chance to ask a few questions. I hope we can talk in more detail later."

Especially if you have not yet submitted for publication the work you are presenting, consider making mote of the questions and comments. Audience members can function as some of your earliest peer reviewers. Keeping their questions in mind when you write may strengthen your paper and hasten its acceptance.

Reference

［1］ Michael J. Moran，Howard N. Shapiro，Daisie D. Boettner，Margaret B. Bailey. Fundamentals of Engineering Thermodynamics，Eighth Edition，Hoboken，New Jersey：Wiley，2014.

［2］ Yunus A. Çengel. Heat transfer：a Practical Approach，Second Edition，New York，N. Y.：McGraw-Hill Companies，Inc.，2003.

［3］ Frank M. White. Fluid mechanics，Seventh Edition，New York，N. Y.：McGraw-Hill Companies，Inc.，2011.

［4］ 2013 ASHRAE Handbook，Fundamentals（I-P Edition），Atlanta，GA：American Society of Heating，Refrigerating and Air-Conditioning Engineers，Inc.，2013.

［5］ 2015 ASHRAE Handbook，Heating，Ventilating，and Air-Conditioning Applications（SI Edition），Atlanta，GA：American Society of Heating，Refrigerating and Air-Conditioning Engineers，Inc.，2015.

［6］ Carter Stanfield，David Skaves. AHRI Fundamentals of HVACR，Second Edition，New Jersey：Pearson Education，Inc.，2013

［7］ Samuel C. Sugarman. HVAC Fundamentals，Third Edition，Indian Trail，Lilburn，Georgia：The Fairmont Press，Inc.，2016.

［8］ Shank K. Wang. Handbook of Air Conditioning and Refrigeration，Second Edition，New York，N. Y.：McGraw-Hill Companies，Inc.，2000.

［9］ Robert A. Day，Barbara Gastel. How to write and publish a scientific paper，6th Edition，Westport，Connecticut，London：Greenwood Press，2006.

［10］ Joshua Schimel. Writing science：how to write papers that get cited and proposals that get funded，Oxford，England，UK：Oxford University Press，2011.

［11］ 吴斐，周频，王军，等. 国际学术交流英语，武汉：武汉大学出版社，2008.

［12］ Cebeci T.，Computational Fluid Dynamics for Engineers，Springer-Verlag，NewYork，2005.

［13］ Tannehill J. D.，Anderson D. A.，Pletcher R. H.，computational Fluid Mechanics and Heat Transfer，2d ed.，Taylor and Francis，Bristol，PA，1997.